SpringerBriefs in Philosophy

SpringerBriefs present concise summaries of cutting-edge research and practical applications across a wide spectrum of fields. Featuring compact volumes of 50 to 125 pages, the series covers a range of content from professional to academic. Typical topics might include:

- A timely report of state-of-the art analytical techniques
- A bridge between new research results, as published in journal articles, and a contextual literature review
- A snapshot of a hot or emerging topic
- An in-depth case study or clinical example
- A presentation of core concepts that students must understand in order to make independent contributions

SpringerBriefs in Philosophy cover a broad range of philosophical fields including: Philosophy of Science, Logic, Non-Western Thinking and Western Philosophy. We also consider biographies, full or partial, of key thinkers and pioneers.

SpringerBriefs are characterized by fast, global electronic dissemination, standard publishing contracts, standardized manuscript preparation and formatting guidelines, and expedited production schedules. Both solicited and unsolicited manuscripts are considered for publication in the SpringerBriefs in Philosophy series. Potential authors are warmly invited to complete and submit the Briefs Author Proposal form. All projects will be submitted to editorial review by external advisors.

SpringerBriefs are characterized by expedited production schedules with the aim for publication 8 to 12 weeks after acceptance and fast, global electronic dissemination through our online platform SpringerLink. The standard concise author contracts guarantee that

- an individual ISBN is assigned to each manuscript
- each manuscript is copyrighted in the name of the author
- the author retains the right to post the pre-publication version on his/her website or that of his/her institution.

More information about this series at http://www.springer.com/series/10082

Kate Chatfield

Traditional and Complementary Medicines: Are they Ethical for Humans, Animals and the Environment?

 Springer

Kate Chatfield
Centre for Professional Ethics
University of Central Lancashire
Preston, UK

ISSN 2211-4548   ISSN 2211-4556   (electronic)
SpringerBriefs in Philosophy
ISBN 978-3-030-05299-7   ISBN 978-3-030-05300-0   (eBook)
https://doi.org/10.1007/978-3-030-05300-0

Library of Congress Control Number: 2018963027

This Springer imprint is published by the registered company Springer Nature Switzerland AG
The registered company address is: Gewerbestrasse 11, 6330 Cham, Switzerland

Foreword

Ethical Issues in Traditional and Complementary Medicines—The Need for a Calm Assessment

Some topics fire argumentative tension like no others. If this tension is expressed through vigorous public debates aiming for a consensus, a major benefit for society can be achieved. If this tension is expressed through frenzied, angry argumentation, it can be harmful to society. Recent examples include the Brexit debate in the UK and the refugee debate in Continental Europe.

A longstanding debate that is partly characterized by angry rather than vigorous discussion is the topic of ethical issues in traditional and complementary medicine (T&CM). For instance, an academic article entitled "Use of Alternative Medicine for Cancer and Its Impact on Survival" (Johnson 2018) is disseminated as "Alternative medicine kills cancer patients" (Pomeroy 2017).

The potential for ethical issues is vast, from profiteering by unscrupulous providers of allegedly helpful complementary therapies to toxic harm through herbs. Yet, each scandal could be matched by multiple numbers of similar scandals in conventional medicine without the same incitement of anger. Profiteering from counterfeit malaria drugs in selected countries in Sub-Saharan Africa alone leads to around 125,000 deaths of children under 5 per year (Renschler 2015). An estimated 66 million potentially clinically significant medication errors occur in the UK each year (Elliot et al. 2018). Hence, it is not as though one side of the debate was free from major ethical challenges.

What is needed is a calm assessment of the issues, informed by considerable expertise and an analytical perspective. This brave assignment is taken on by Kate Chatfield and taken on successfully. Chatfield untangles the web of angry and reasoned positions on ethical issues in T&CM and comes to valuable conclusions.

Conventional medicine and T&CM can both be practised ethically and unethically. They are both responsible for major adverse effects in humans, detrimental impacts upon the environment, as well as harm and suffering for animals, where conventional medicine fares much better than T&CM is on regulation. It is much more difficult to be an unethical conventional health practitioner than to be an unethical T&CM practitioner. That is because regulation helps ensure ethical conduct. Chatfield therefore appeals to regulators to close gaps. For instance, the availability of unlicensed, unregulated and potentially lethal T&CM medications via the Internet requires global attention and cooperation.

At the same time, some comparisons weigh in favour of T&CM, which fares much better than conventional health care on environmental impact. As Chatfield shows, the current provision of conventional medicine is vastly more damaging on the planet than the provision of most complementary medicines in terms of energy use, reliance upon oil and pollution of the environment.

In the conclusion of her book, Chatfield calls for a resilience toolkit. When health is seen as an ability to adapt to physical, mental and social challenges, in other words, resilience, prevention becomes an imperative. To leave T&CM outside of this toolkit, in a world where, for many, it serves as the only form accessible health care, would be unethical.

Preston, UK   Doris Schroeder
                                                         Professor of Moral Philosophy
                                                      University of Central Lancashire

References

Elliott R, Camacho E, Campbell F, Jankovi, D, St James MM, Kaltenthaler E, Wong R, Sculpher M, Faria R (2018) Prevalence and economic burden of medication errors in the NHS in England. Policy Research Unit in Economic Evaluation of Health & Care Interventions (EEPRU)

Johnson SB, Park HS, Gross CP, Yu JB (2018) Use of alternative medicine for cancer and its impact on survival. J Nat Cancer Inst 110(1)

Pomeroy R (2017) Alternative medicine kills cancer patients, study finds. RealClear Science. https://www.realclearscience.com/quick_and_clear_science/2017/08/14/alternative_medicine_kills_cancer_patients_study_finds.html. Accessed 16 Oct 2018

Renschler JP, Walters KM, Newton PN, Laxminarayan R (2015) Estimated under-five deaths associated with poor-quality antimalarials in sub-Saharan Africa. Am J Trop Med Hyg 92 (6):119–126

Contents

Chapter 1
Defining Traditional and Complementary Medicine

Abstract 'Complementary', 'alternative', 'traditional' or even 'pseudomedicine', are just some of the words that are used to describe this body of therapeutic interventions. Collectively, these terms evoke an array of seemingly disparate connotations, indicative of a wide range of perspectives. Indeed, opinions about their worth span a full spectrum from: 'essential and highly valued forms of health care', to: 'no better than placebos that are proffered by charlatans'. This polarisation of perspectives is a significant challenge when it comes to explanation of what is meant by 'complementary' or 'traditional' medicine, as is the broad range of individual interventions that fall under this sizeable umbrella. The challenges for establishing a clear definition are explained and arguments given for adopting the World Health Organization's definition of 'traditional and complementary medicine' (T&CM). An overview of the controversial nature of the subject matter provides a backdrop to subsequent ethical analysis in later chapters.

Keywords Complementary medicine · Traditional medicine
Alternative medicine · Evidence-based medicine · Definition · Benefits
Controversy

1.1 What's in a Name?

Names are an important key to what a society values. Anthropologists recognize naming as one of the chief methods for imposing order on perception.

David S. Slawson

The field of traditional and complementary medicine (T&CM) comprises a broad range of disciplines with varied medical philosophies, diagnostic methods, therapeutic interventions, and life style approaches that are grouped collectively because they are in some way dissimilar to health care that is offered by the prevailing healthcare system (Sharma 1992). The precise nature of the dissimilarity is impossible to specify because the range of procedures offered both within and outside of conventional healthcare systems is extremely diverse. Nevertheless,

K. Chatfield, *Traditional and Complementary Medicines: Are they Ethical for Humans, Animals and the Environment?* SpringerBriefs in Philosophy
https://doi.org/10.1007/978-3-030-05300-0_1

1

many bodies have attempted to define and delineate this group of therapies. For example, *Cochrane Complementary Medicine*, the Cochrane field that coordinates the T&CM-related activities, applied their own criteria to identify fifty-one different therapies that are used in treating or preventing disease[1] (Wieland et al. 2011). T&CM therapies may differ from conventional biomedicine, not only in their methods, but also in some of their underlying philosophies (Cassidy 2002). The way that many of these disciplines define health, illness and the healing process can depart significantly from that of conventional medicine (Tataryn 2002).

The problematic nature with formulating a definition is exacerbated by the use of different names for this group of interventions which, in themselves, evoke specific connotations. For instance, names such as 'complementary', 'alternative', 'integrative', and 'holistic', convey particular viewpoints about the manner in which the therapy is used: 'traditional' about the historical roots of the therapy; 'natural' about the source materials or a self-healing response; while some names, such as 'quackery' and 'pseudomedicine' are clearly intended as derogatory. The names applied to this group of therapies reveal inherent preconceptions and prejudices: "Alternative medicine is nothing more than a label; an abstract socio-cultural construct that serves the establishment and much less so the patient" (Caspi et al. 2003).

The term 'alternative medicine' was first popularised in the United States and Europe in the 1970s when it was noted that some people, when given the choice, were electing to use certain unconventional forms of health care, also then termed 'irregular medicine' or 'fringe medicine' (Fulder 1988). In the late 1980s the expression 'complementary medicine' was introduced in an attempt to present a more accurate reflection of the way in which most people use these therapies as an adjunct to conventional treatment (BMA 1993). Currently, in most English speaking countries, the words 'complementary' and 'alternative' are used interchangeably to describe this body of therapeutic systems (National Health Service 2018). This is frequently designated by the acronym 'CAM' (complementary *and* alternative

---

[1] Acupressure, Acupuncture, Alexander technique, Aromatherapy, Arts therapy (eg, dance therapy, drama therapy, music therapy), Ayurvedic traditional medicine, Balneotherapy (natural spring water bathing), Bee products(e.g. honey, propolis, royal jelly), Biofeedback, Chelation therapy (removal of toxic heavy metals from the body), Chinese traditional medicine, Chiropractic, Colour therapy, Craniosacral manipulation, Dietary supplements, Diet therapy, Distant healing, Electric stimulation therapy (e.g. TENS machine), Electromagnetic therapy (magnets), Eye Movement Desensitization and Reprocessing (EMDR) (a form of psychotherapy), Feldenkrais method (awareness through movement), Herbal supplements, Homeopathy, Hydrotherapy, Hyperbaric oxygenation, Hypnosis, Imagery, Light therapy, Magnetic field therapy, Massage, Meditation, Morita therapy (a Japanese mindfulness technique), Moxibustion, Naturopathy, Osteopathic manipulation, Ozone therapy, Play therapy, Prolotherapy (dextrose injections for non-surgical ligament reconstruction), Qigong, Reflexology, Reiki therapy, Relaxation techniques, Snoezelen (controlled multisensory environment most often used for people with learning difficulties), Speleotherapy (exposure to salt air), Spiritual healing, Tai chi, Therapeutic touch, Traditional healers and healing practices (other than Chinese), Tui na (Chinese manipulative therapy), Ultrasonic therapy (using sound waves to penetrate soft tissues), Yoga.

medicine), denoting usage either as an alternative to conventional treatment, or in a complementary manner. Globally the term most often used is 'traditional medicine', denoting the cultural heritage of these forms of medicine in certain parts of the world.

1.2 The Differences Between T&CMs and Conventional Medicine

Many have proposed definitions for T&CM; some by attempting to identify the essential qualities of the therapies but, more commonly, by seeking to demarcate T&CMs from conventional medicine. Back in 1995, the following definition was proposed by a working group from what was then entitled the *Office of Alternative Medicine* (OAM):

> Complementary and alternative medicine (CAM) is a broad domain of healing resources that encompasses all health systems, modalities, and practices and their accompanying theories and beliefs, other than those intrinsic to the politically dominant health system of a particular society or culture in a given historical period. CAM includes all such practices and ideas self-defined by their users as preventing or treating illness or promoting health and well-being (OAM 1997).

According to this definition, wherever the politically dominant or conventional health system is biomedicine then CAMs comprise the broad domain of healing resources other than those intrinsic to biomedicine. If, in another country, biomedicine is not the politically dominant health system then the range of practices that are considered complementary or alternative will be different. Eskinazi, ascribes the demarcation to the philosophically challenging nature of CAM approaches, suggesting that the definition should be:

> A set of healthcare practices (i.e. already available to the public) that are not readily integrated into the dominant healthcare model because they pose challenges to diverse societal beliefs and practices (cultural, economic, scientific, medical, and educational) (Eskinazi 1998).

While Benitez-Bribiesca distances such practices from science:

> Health care services around the world can be roughly classified into two groups according to their medical orientation: scientific medicine and 'traditional medicine' (Benitez-Bribiesca 2000).

The above definitions define CAM approaches in the negative, by telling us something about what it is *not*, rather than what it is. The terms 'alternative' and 'complementary' imply an extension; a phenomenon is not alternative or complementary in itself, only in relation to something else. However, any medicine may be 'conventional' or 'standard' in its own setting. Traditional Chinese medicine is perceived as conventional in China, and Ayurvedic medicine as conventional in India, whereas both are considered as complementary or alternative in Western

Table 1.1 Comparison of T&CM and conventional medical approaches

Traditional and complementary medicine	Conventional medicine
Takes a holistic approach to diagnosis and treatment; i.e. physical, mental, spiritual, and environmental factors are considered (Barrett et al. 2003)	Takes a reductionist approach to diagnosis and treatment; i.e. various specialists are concerned with individual illnesses (Milgrom 2006)
Commonly invokes an underlying vitalistic[a] doctrine (Bellavite 2003)	Based upon a biomedical understanding of the body (Colquhoun and Isbell 2007)
Treatment is tailored to the individual (Franzel et al. 2013)	Treatments are largely based upon what works for the 'average' person (Cronje and Fullan 2003)
Treatments are derived from, and in tune with, nature (Ostendorf 1991)	Treatments can be highly medicalised (Hughes 2008)
Cure is encouraged from within; there is high regard for self-healing (Fulder 1988)	Cure is interventionist in approach; symptoms are controlled (di Sarsina et al. 2012)
Commonly used to promote and maintain health, for preventative purposes (Patriani Justo and dé Andrea Gomes 2008)	Most commonly used in a reactive manner for treating existing complaints (Teixeira 2009)
High regard for autonomy; patient is active partner in the healing process (Sharma 1992)	Often criticised for taking a paternalistic approach; doctors are the experts (White 2000)

[a]*Vitalism* is the metaphysical doctrine that living organisms possess a non-physical inner force or energy that imparts life

countries. Additionally, the image of biomedicine as a monolithic structure is inaccurate: its practice is deeply affected by cultural norms and values around the world (Stein 1990). Cassidy asserts that, if we are agreed that there are many medicines globally, with a wide range of philosophical underpinnings, cultural connotations and varying degrees of prominence, then we must come to see that *all* are alternatives, and *all* can complement others. In short, conventional medicine ought not to be treated as the standard against which all others are compared but as one among many, itself a CAM practice (Cassidy 2002).

While it is true that conventional biomedicine might itself be viewed as CAM in some cultures/scenarios, there does appear to be some common underlying assumptions that apply to most, if not all, T&CM therapies and which differentiate them from a conventional biomedical approach, at least in the Western world. Some of the most widely accredited differences in perspective are summarised in Table 1.1.

Listed here are common differences in approach that are often cited in CAM literature.

T&CMs have hugely diverse backgrounds, but there are some qualities that most seem to share. However, there are no specific qualities that apply exclusively and equally to all subsets. For example, a fundamental similarity between most T&CMs is that they appear to apply a non-Cartesian view of health, which makes little distinction between the body, emotions, mind or spirit as separate sources of disease. All dimensions are seen to have an impact upon health and symptoms of ill health can be expressed on any level. Nevertheless, a spectrum exists between reductionism and holism in both conventional medicine and T&CM and both can be said to span the spectrum (White 2000). Many conventional medical practitioners take a holistic approach to treatment and many T&CM treatments are aimed at a specific ailment, such as the use of natural antibacterial and anti-viral supplements, which are aimed directly at microorganisms, just as their conventional counterparts are.

Another quality shared by many T&CM interventions is that of taking an individualistic approach to treatment, such that patients receive treatments that are tailored to their specific needs. Again, this is not common to all; there are many examples of a specific therapeutic use of treatments such as the use of the herb St John's Wort for anxiety and depression, and the use of the homeopathic remedy Arnica for bruising. Furthermore, with the onset of personalised medicine and the use of individual genetic profiles to determine treatment, conventional medicine can also be individualised.

Most T&CMs are credited with being 'natural medicines' involving the therapeutic use of natural methods and materials as opposed to human-made or chemically synthesised products. However, the claim of 'natural' is not exclusive to T&CM; many conventional medications stem from nature, such as quinine and artemisimin, both sources of anti-malarial drugs.

1.3 A Pragmatic Approach

In order to circumnavigate the problems with defining T&CM, some institutions, like *Cochrane Complementary Medicine*, have taken a pragmatic approach and developed their own operational definitions. There is a practical need to decide upon what can be considered T&CM for the purpose of treatment classification, health expenditure and so on. People creating databases must decide which treatments to store information about and how to categorise this information within subsets. Practical decisions need to be made about what is included in the education and training of health professionals that, to some degree, depends upon what falls within and what lies outside the domain of conventional medicine.

While attempting to classify their reviews as T&CM medicine-related, many challenges were encountered by *Cochrane Complementary Medicine* because the available theoretical definitions did not help in the operational activity of deciding which to include (Wieland et al. 2011). The major problem the Cochrane team perceived, when attempting to operationalise a definition that relies upon contrast with conventional medicine, is that the conventional medical model changes over time. Instead, they proposed three specific questions to help define the parameters of T&CM approaches:

(1) Is the historical notion of the therapy complementary or conventional?
(2) Is use of the therapy for a particular condition currently considered to be a standard treatment within the conventional medical system, and
(3) In what setting is the therapy delivered, and by whom?

It is through application of these criteria, that they identified the fifty-one therapies previously mentioned. Cochrane are careful to highlight that they do not consider their definition as definitive. Indeed, they question whether it is possible to identify a definitive set of therapies that are universally agreed upon.

1.4   The T&CM Family

The task of generating a specific definition is most likely to continue to evade consensus; the T&CM community has been struggling for many years to come up with a single definition agreed by all, with no success (McIntyre 2001). Perhaps one of the main challenges in identifying a precise and universally agreed view of what T&CM is, even among those desiring to pursue common goals, is that people may not be 'speaking the same language' or sharing the same meanings for terms they use, such as 'traditional', 'complementary', 'alternative' and 'medicine'. It has been suggested that it is difficult to forge appropriate language for definition and description in a field as contested, politically charged, and value-laden as this, because these linguistic acts presuppose a particular point of view, and often carry moral tone (O'Connor 1995). They may suggest, by their selection of words, whether the subject is to be regarded favourably or unfavourably (OAM 1997). However, in everyday life, people do refer to *traditional* or *complementary* medicines as a matter of routine and therefore it seems reasonable to postulate that there is some kind of shared understanding of the terms. On the one hand we appear to have a shared understanding of what they mean and, on the other, no defining quality that applies to all types. One solution to this apparent conundrum can be found in Ludwig Wittgenstein's concept of 'family resemblance'.

1.4.1   A Family Resemblance

The notion that a definition must, by its nature and purpose, apply fully to all members of the class to which it applies was challenged by Ludwig Wittgenstein who pointed instead to 'family resemblance' as a more suitable analogy for ways in which we connect particular uses of the same word. Wittgenstein introduces the concept of family resemblances in his criticism of the traditional idea that all entities that fall under a given term must have the same set of properties or features in common. According to Wittgenstein, there is no reason to look for one essential core quality in which the meaning is located, but rather that we should think of a word's uses as 'a complicated network of similarities overlapping and criss-crossing' (Wittgenstein 1967). To illustrate his point, Wittgenstein described the use of the word 'game':

Consider for example the proceedings that we call 'games'. I mean board-games, card-games, ball-games, Olympic games, and so on. What is common to them all? Don't say: 'There must be something common, or they would not be called 'games'' -but look and see whether there is anything common to all. For if you look at them you will not see something that is common to all, but similarities, relationships, and a whole series of them at that (Wittgenstein 1967, p. 66).

Wittgenstein's argument is that the entities under a given term need not have any one thing in common. The similarities that exist between them are like the resemblances between members of a family: build, features, colour of eyes, gait, temperament, and so on. These properties may overlap and criss-cross, but there is no one thing that is exactly the same for them all. In this way 'games' form a family and, accordingly, so would 'T&CM therapies'. Individual therapies may be related to each other in ways that overlap, some of which have already been mentioned; natural, holistic, individualised, promoting self-healing, and so on. The idea that there may be a family resemblance between T&CMs helps to explain how it is possible to have a shared understanding of the term, but the boundaries between T&CM and conventional medicine may still be somewhat blurred.

As previously stated, what is considered complementary or alternative in some localities may be considered as part of the established healthcare system in others. While biomedicine dominates medical healthcare in the Western world, this is not the case globally.

1.5   A Global Perspective

Usage of treatments other than conventional biomedicine may be as high as 80% of the population in some parts of Asia and Africa where many people rely upon other forms of medicine for their primary health care (WHO 2013). For people in low and middle income countries (LMICs) it is often the case that appropriate conventional medicine is cost-prohibitive, inaccessible, or simply not available. In such circumstances, people are largely dependent on some form of local or traditional interventions, utilising specific treatments that have a cultural connection, and are readily available and affordable.

For instance, in Africa the majority of medical doctors are situated within urban areas and cities, inaccessible for those living in rural areas. Moreover, the ratio of traditional health practitioners to population in Africa is 1:500 compared to a ratio of 1:40,000 for medical doctors to population (Abdullahi 2011). In addition to being cheaper and more accessible, there may be a distinct preference for indigenous forms of health care. For instance, traditional African healers may be sought because they offer treatments for the more intangible necessities of health and wellbeing that are not part of the conventional medic's toolkit, such as the warding off of evil, bringing luck and good fortune, obeying instructions from ancestors, and the interpretation of dreams (Omonzejele and Maduka 2011).

In China, Korea and Japan, East Asian medicine predates conventional medicine by several thousand years, is highly valued, and is utilised by between 40 and 76% of the populations. In these three countries, the East Asian medicine methods of acupuncture, herbal medicine, moxibustion, cupping and manual therapies are integrated into the national health care systems in a manner that is respectful of the historical and cultural background and allows people relatively equal access to both forms of medicine (Park et al. 2012). Clearly, in these circumstances, the designation of 'complementary' or 'alternative' would not be appropriate for the types of intervention that are not intrinsic to biomedicine.

Unlike most other attempts at classification, the World Health Organization (WHO) distinguish between healthcare interventions that are used in a complementary or alternative manner, away from their geographical origins, and those that are rooted in local tradition and culture (traditional medicine) in their definition as follows:

Traditional Medicine

Traditional medicine is the sum total of the knowledge, skills, and practices based on the theories, beliefs, and experiences indigenous to different cultures, whether explicable or not, used in the maintenance of health as well as in the prevention, diagnosis, improvement or treatment of physical and mental illness.

Complementary/Alternative Medicine

The terms 'complementary medicine' or 'alternative medicine' are used interchangeably with traditional medicine in some countries. They refer to a broad set of health care practices that are not part of that country's own tradition and are not integrated into the dominant health care system (WHO 2000).

Acording to the WHO, any form of healthcare that is indigenous to a particular locality can be classified as traditional medicine within that locality but, outside that locality, may be considered as complementary or alternative. Thus, acupuncture and the use of Chinese herbal medicines should be classified as traditional medicine when used within China, but as complementary or alternative medicine elsewhere in the world; what is considered traditional and what is considered complementary or alternative will therefore vary from country to country and from region to region.

For the sake of clarity, there is value in drawing a distinction between traditional or indigenous forms of healing that are rooted in local culture and tradition and those which are used, as they commonly are in the Western world, in a complementary or alternative manner; it is important to acknowledge that there may be different motivations for use, different levels of reliance, and different degrees of autonomy involved in choice of healthcare. Hence, I am adopting the WHO's distinction in this book.

1.6   T&CM Usage

The wide scale use of T&CM, and associated practical and ethical concerns, are the focus of attention for the WHO in their traditional medicine strategy 2014–2023 (WHO 2013). According to the WHO, different forms of T&CM are found in almost every country in the world and the demand is increasing (ibid.). Precise figures for global usage are elusive as demand differs greatly between regions, socio-economic groups, cultures and health complaints, even within individual countries. However, estimates suggest that usage is highest in parts of Asia and Africa but elsewhere in the world there is evidence to suggest that demand for T&CM is high and increasing for certain types, in spite of the ready availability of conventional care. A systematic review of usage in Europe included studies that had found between 5 and 74.8% in different countries. Excluding any form of spiritual prayer, the data demonstrated that, in Europe, chiropractic manipulation, herbal medicine, massage, and homeopathy were the therapies most commonly used by the general population (Frass et al. 2012). In the USA, overall, 34% of adults used any form of T&CM in 2012 with dietary supplements (other than vitamin and mineral) being the most commonly utilised (Clarke et al. 2015). According to Reid et al. (2016), T&CM use is substantial across contemporary Australia and the increasing use of T&CM services by the general population has resulted in T&CM becoming an important subject amongst Australian primary health care professionals and policy makers. Overall, it has been estimated that as many as one in two adults in Australia and Europe has used some form of T&CM and this figure has remained relatively constant in surveys over recent years (Harris et al. 2012).

Many reasons have been cited for the popularity of T&CM such as dissatisfaction with conventional medicine, a lack of effective conventional medication, and the presence of unwanted side effects from conventional care. However, most research evidence suggests that T&CM use can best be understood as a component of self-care management in general, and not as a rejection of conventional medicine (Bishop et al. 2010). For example, a person with a chronic complaint such as back pain might choose between an acupuncturist, osteopath or chiropractor, or try a range of different products, such as herbal tinctures or topical ointments, alongside conventional medications. There is evidence to suggest that T&CM interventions are being used in this way for treatment of a broad spectrum of both chronic and acute health complaints including cancer, complaints related to pregnancy and childbirth, mental health problems, back pain, addiction, arthritis, allergies, and a range of childhood complaints.

1.7   T&CM Benefits

Aside from any potential direct therapeutic benefits of T&CMs there are also broader benefits.

Herbal medicines, in particular, contribute directly to the development of conventional medications with as many as one-third to one-half of pharmaceutical drugs being derived from plants (Abbott 2014). These include many widely used forms of medicine, for example:

- Taxol, an antitumor agent that is derived from *Taxus brevifolia* (Pacific yew).
- Morphine and codeine, widely used analgesics that are derived from *Papaver somniferum* (poppy).
- Digoxin, a cardiotonic that is derived from *Digitalis purpurea* (purple or common foxglove).
- Quinine, an antimalarial drug that is derived from *Cinchona ledgeriana* (sometimes called the quinine tree).

Economic benefits are significant for some countries where T&CM is an important source of income for many individuals and communities. According to the WHO (2013), the T&CM sector now plays a significant role in the economic development of a number of countries. This is highly evident in China, for example, where global promotion of traditional Chinese medicine has been used to foster economic development, generating billions of United States dollars in revenue annually (Abbott 2014).

The value of the contribution that T&CM makes to global health should not be underestimated, as emphasised by the WHO Director-General, Dr. Margaret Chan, who, when addressing the International Conference on Traditional Medicine for South-East Asian Countries in 2013, declared that:

> Traditional medicines, of proven quality, safety, and efficacy, contribute to the goal of ensuring that all people have access to care. For many millions of people, herbal medicines, traditional treatments, and traditional practitioners are the main source of health care, and sometimes the only source of care. This is care that is close to homes, accessible and affordable. It is also culturally acceptable and trusted by large numbers of people. The affordability of most traditional medicines makes them all the more attractive at a time of soaring health-care costs and nearly universal austerity (As cited in WHO 2013).

A further noteworthy benefit, highlighted by the WHO (2013), is that T&CMs have a function in disease prevention and can be used to enhance and maintain health. Appropriate use has the potential to make T&CMs highly cost-effective, helping to reduce pressure on health care systems by reducing overall costs. This is a particularly vital consideration given burgeoning healthcare costs and the growing need for development of sustainable forms of health care.

In spite of the potential for benefit, its widespread usage, and the apparent acceptance of T&CM by the general public, the topic of T&CM can lead to emotionally charged debates. In some parts of the world there are particularly vocal objectors to the usage of T&CM and it has variously been described as a form of 'medical heresy' (Stambolovic 1996), no better than 'superstition' (Singh and Ernst 2008), and 'pseudoscientific' (Hall 2009). In an ethical analysis, the impetus for such fervent contempt of the field is worthy of deliberation.

1.8   T&CM Controversy

The primary objection to use of T&CM interventions focuses upon a lack of a rigorous evidence base for most types of T&CM, as is required for conventional medical health care (Ernst and Smith 2018). The question asked is certainly reasonable: *Is it ethical to prescribe treatments that have no scientific evidence of efficacy or safety?* No new conventional medications are introduced without rigorous testing through established scientific methods such as clinical trials.

The practice of T&CM, however, is primarily rooted in case-based, empirical evidence, passed down through generations. This form of knowledge, whether in written or purely verbal form, is subject to testing for effectiveness and safety only when applied in the real world to individuals. For some types of T&CM, such as Ayurveda, such application has taken place over many centuries.

This type of historical evidence does not satisfy those who demand that all healthcare interventions should be rooted in a scientific evidence base. In Western healthcare the introduction of evidence-based medicine (EBM) has been heralded as the most recent revolutionary phase, described as a 'paradigm shift' that will change medical practice for years ahead (Guyatt and Rennie 2001). This movement was initiated in 1992 when a group of physician-researchers, known as the Evidence-Based Medicine Working Group (EBMWG 1992) published an article urging physicians to base clinical decisions on reliable evidence. In this way variable, intuitive judgements would be replaced with rational calculation and the use of research.

The notion of EBM is predicated upon the assumption that there is a hierarchy of evidence for medical interventions, such that certain types of evidence are deemed of greater value than others. At the top of this hierarchy are placed results from randomised controlled trials (RCTs) and systematic reviews[2] or meta-analyses[3] of these trials. The synthesis of large amounts of clinical trial data into manageable systematic reviews or meta-analyses is meant to revolutionise medical practice and offer objective and politically transparent criteria for treatment choice and funding decisions. Proponents of EBM commend themselves on their attempt to 'realign medicine with science', while critics object that EBM emphasises exclusively the science of medicine while denying the art of medical practice (Miettinen 2001). Since the broad adoption of EBM principles, there has been increasing pressure on all forms of T&CM to provide the kind of evidence that is deemed of high value, from RCTs and reviews of these trials, to enable comparative assessment of efficacy and safety. Without this type of evidence, most governments are unwilling to fund provision, research or development of T&CM on anything but a small scale.

Attempts have been made to address this requirement for evidence but subsequently many T&CM researchers have described reasons why this is proving to be a challenging task (Bell et al. 2012). For example, for T&CM interventions, such as acupuncture and manipulative therapies, it is impossible to create a convincing and

---

[2] A systematic review is a means of examining results from more than one trial to look for trends.

[3] A meta-analysis takes an overall view further by performing statistical analysis on the combined results to look at the statistical significance of trends.

yet inactive placebo treatment for comparison in an RCT. This type of research relies upon the participants in the research trial not knowing whether they are receiving active treatment or placebo. While efforts have been made: acupuncture has been tested against 'sham acupuncture' (Kim et al. 2011) and chiropractic against 'sham chiropractic' (Hannah et al. 2012), results from many of these trials indicate that 'sham' treatments often have an effect themselves and therefore cannot be considered effective placebos (Witt and Schützler 2012).

Aside from many and varied methodological challenges, the greatest challenge for T&CM research, seems to be the scarcity of research funding. Government, charitable and private funding for research into T&CM is low, with the result that research funding for T&CM is vastly out of proportion to the prevalence of T&CM use. This can be seen clearly in the United States where funding for National Centre for Complementary and Integrative Health (NCCIH) activity in 2015 amounted to $124 million while total funding of health research and development activity in the United States amounted to $158,716 million (Research America 2016) indicating that NCCIH funding is equivalent to less than 0.08% of all medical research funding. Funding for wide scale T&CM research, of the type that would be needed to develop a robust evidence base, is simply not available.

Some argue that it would be unethical to provide funds for investigation of products and services deemed *implausible*, as funding research under these circumstances would constitute a waste of resources (Shaw 2011). This leads to a stalemate position on the issue of evidence in T&CM; on the one hand T&CM is criticised for not having a robust evidence base and on the other hand T&CM research is not being funded to enable such research to be performed.

These challenges, together with many others, have contributed to the current situation where there is considerable debate about whether or not T&CM interventions are, or should be, an important element of health care. Argument is ongoing about what constitutes reliable evidence on this topic and how that evidence might be assembled, as can be seen in Table 1.2, which summarises the contrasting polarities of opinion in the debate focusing on evidence.

Table 1.2 Contrasting opinions on evidence for effectiveness of T&CM

T&CM is effective	T&CM is not effective
Historical and case-based evidence demonstrates clinical effectiveness	Historical and case-based evidence does not demonstrate causal relationships and hence is not acceptable as proof of efficacy
People would not continue to use T&CM interventions if they were ineffective	Perceived benefits may stem from placebo effects
Research methods most highly placed in the evidence hierarchy of EBM (RCTs and systematic reviews) do not adapt well to complex T&CM interventions	The only acceptable test of efficacy is through a randomised placebo-controlled trial
There is a distinct lack of funding available for T&CM research	Resources are better focussed on conventional medical research

Distinct polarities of belief leading to a stalemate position on the issue of evidence in T&CM.

Clearly, the above two positions cannot be reconciled easily and the debate about efficacy remains unresolved in most areas of T&CM. Nonetheless, if it is to be deemed unethical to use medicines and interventions that do not have a robust research evidence base then much of conventional medicine would have to be stopped. Estimates vary widely (from 11 to 70%) as to the exact proportion of conventional interventions that are robustly evidence-based (Oliver 2014). In the US, it has been estimated that around half (51%) of recommendations in primary care are based upon good or fair quality patient-oriented evidence, but only 18% are based on patient-oriented evidence from consistent, high-quality studies (Ebell et al. 2017). It is obviously of ethical concern that healthcare interventions are effective, but it is beyond the scope of this analysis to make judgements about efficacy in T&CM. Nevertheless, since the practice of T&CM (and many conventional healthcare interventions) continues in spite of scepticism about efficacy, ethical challenges about the practice, production, promotion and usage of T&CM require attention.

Given that much of T&CM is under-researched, not only in terms of efficacy but also in terms of safety and cost-effectiveness, this inevitably generates ethical concerns. In addition, the practice of T&CM in many countries has largely existed as an independent, parallel and disparate healthcare system (Kerridge and McPhee 2004) and professional standards are extremely variable (Tyreman 2011). While many countries are attempting to develop policy for, and regulation of, T&CM, practices are inconsistent and vary from country to country. In some countries T&CM remains completely unregulated (WHO 2013). As the complexities of this situation become better understood, specific ethical challenges and legal issues are increasingly highlighted (Smith et al. 2016).

Specific ethical challenges to T&CM are numerous and include issues such as safety of interventions, quality control of products, regulation and competence of practitioners, fully informed consent, the giving of misleading information, and concerns that T&CM treatment may delay or prevent patients from seeking mainstream health care such that 'the harm done by omitting evidence-based medical treatment is potentially significant' (Shaw 2010). These ethical issues clearly call into question the potential for benefit and how to avoid harm, which in turn must be considered in the light of an individual's right to use their health care of choice. In addition, there are ethical considerations that extend beyond the autonomy and the wellbeing of patients and practitioners to embrace issues such as fairness and equity, and sustainability.

This enquiry begins with the underlying assumption that there are ethical challenges for the use and practice of T&CM, just as there are for the use and practice of conventional medicine. In subsequent chapters, I will identify the primary ethical challenges for T&CM and analyse the most prominent for humans, animals and the environment. Where appropriate, my analysis will include comparisons with conventional medicine. Before undertaking the analysis of ethical issues associated with T&CM, in my next chapter I explain what I mean by 'health' as this, presumably, is the ultimate goal for all health care interventions.

References

Abbott R (2014) Documenting traditional medical knowledge. World Intellectual Property Organization. http://www.wipo.int/export/sites/www/tk/en/resources/pdf/medical_tk.pdf. Accessed 4 Apr 2018

Abdullahi AA (2011) Trends and challenges of traditional medicine in Africa. Afr J Tradit Complement Altern Med 8(5S):115–123

Barrett B, Marchand L, Scheder J, Plane MB, Maberry R, Appelbaum D, Rakel D, Rabago D (2003) Themes of holism, empowerment, access, and legitimacy define complementary, alternative, and integrative medicine in relation to conventional biomedicine. J Altern Complement Med 9(6):937–947

Bell IR, Koithan M, Pincus D (2012) Methodological implications of nonlinear dynamical systems models for whole systems of complementary and alternative medicine. Complement Med Res 19(1):15–21

Bellavite P (2003) Complexity science and homeopathy: a synthetic overview. Homeopathy 92(4):203–212

Benitez-Bribiesca L (2000) Alternative medicine: to teach or not to teach. Arch Med Res 31(6):537–538

Bishop FL, Yardley L, Lewith GT (2010) Why consumers maintain complementary and alternative medicine use: a qualitative study. J Altern Complement Med 16(2):175–182

British Medical Association (1993) Complementary medicine: new approaches to good practice. Oxford University Press, Oxford

Caspi O, Sechrest L, Pitluk HC, Marshall CL, Bell IR, Nichter M (2003) On the definition of complementary, alternative, and integrative medicine: societal mega-stereotypes vs. the patients' perspectives. Altern Ther Health Med 9(6):58–62

Cassidy CM (2002) Commentary on terminology and therapeutic principles: challenges in classifying complementary and alternative medicine practices. J Altern Complement Med 8(6):893–895

Clarke TC, Black LI, Stussman BJ, Barnes PM, Nahin RL (2015) Trends in the use of complementary health approaches among adults: United States, 2002–2012. Natl Health Stat Rep 79:1

Colquhoun D, Isbell B (2007) Credible endeavour or pseudoscience? Times Higher Education Supplement, 6 April , p 14

Cronje R, Fullan A (2003) Evidence-based medicine: toward a new definition of 'rational' medicine. Health: Interdiscip J Soc Study Health, Illn Med 7(3):353–369

di Sarsina PR, Alivia M, Guadagni P (2012) Widening the paradigm in medicine and health: person-centred medicine as the common ground of traditional, complementary, alternative and non-conventional medicine. In: Costigliola V (ed) Healthcare overview. Advances in predictive, preventive and personalised medicine. Springer, Dordrecht

Ebell MH, Sokol R, Lee A, Simons C, Early J (2017) How good is the evidence to support primary care practice? Evid Based Med 22(3):88–92

Ernst E, Smith K (2018) More harm than good?: the moral maze of complementary and alternative medicine. Springer International Publishing, Basel

Eskinazi D (1998) Factors that shape alternative medicine. J Am Med Assoc 280(18):1621–1623

Evidence Based Medicine Working Group (1992) Evidence-based medicine: a new approach to teaching the practice of medicine. J Am Med Assoc 268(17):2420–2425

Franzel B, Schwiegershausen M, Heusser P, Berger B (2013) Individualised medicine from the perspectives of patients using complementary therapies: a meta-ethnography approach. BMC Complement Altern Med 13(1):124

Frass M, Strassl RP, Friehs H, Müllner M, Kundi M, Kaye AD (2012) Use and acceptance of complementary and alternative medicine among the general population and medical personnel: a systematic review. Ochsner J 12(1):45–56

Fulder S (1988) The handbook of complementary medicine, 2nd edn. Open University Press, Oxford

Guyatt GH, Rennie D (2001) Users' guides to the medical literature: a manual for evidence-based clinical practice. American Medical Association, Chicago

Hall H (2009) Homeopathy: still crazy after all these years. Skeptic 15(1):8–9

Hannah L, Ross GJ, Oliver PT (2012) Pilot Study: The suitability of sham treatments for use as placebo controls in trials of spinal manipulative therapy: a pilot study. J Bodyw Mov Ther. https://doi.org/10.1016/j.jbmt.2012.06.005

Harris PE, Cooper KL, Relton C, Thomas KJ (2012) Prevalence of complementary and alternative medicine (CAM) use by the general population: a systematic review and update. Int J Clin Pract 66(10):924–939

Hughes BM (2008) How should clinical psychologists approach complementary and alternative medicine? Empirical, epistemological, and ethical considerations. Clin Psychol Rev 28(4):657–675

Kerridge IH, McPhee JR (2004) Ethical and legal issues at the interface of complementary and conventional medicine. Med J Aust 181:164–167

Kim DI, Jeong JC, Kim KH, Rho JJ, Choi MS, Yoon SH, Choi S-M, Kang KW, Ahn HY, Lee MS (2011) Acupuncture for hot flushes in perimenopausal and postmenopausal women: a randomised, sham-controlled trial. Acupunct Med 29(4):249–256

McIntyre M (2001) The House of Lords select Committee's report on CAM. J Altern Complement Med 7(1):9–11

Miettinen OS (2001) The modern scientific physician: 1. Can practice be science? Can Med Assoc J 165(4):441–442

Milgrom LR (2006) Is homeopathy possible? J R Soc Promot Health 126(5):211–218

National Health Service (2018) Complementary and alternative medicine. https://www.nhs.uk/conditions/complementary-and-alternative-medicine/. Accessed 10 Apr 2018

O'Connor BB (1995) Healing traditions: alternative medicine and the health professions. Studies in health, illness, and caregiving, vol xxiv. Univeristy of Pennsylvania Publishing, Philadelphia

Office of Alternative Medicine (1997) Defining and describing complementary and alternative medicine. In Panel on definition and description, CAM research methodology conference, April 1995. Altern Ther Health Med 3(2):49–57

Oliver D (2014) Evidence based medicine needs to be more pragmatic. BMJ 349(7966):g4453

Omonzejele PF, Maduka C (2011) Metaphysical and value underpinnings of traditional medicine in West Africa. Chin J Integr Med 17(2):99–104

Ostendorf GM (1991) Naturopathy and alternative medicine—definition of the concept and delineation. Das Offentliche Gesundheitswesen 53(2):84–87

Park H-L, Lee H-S, Shin B-C, Liu J-P, Shang Q, Yamashita H, Lim B (2012) Traditional medicine in China, Korea, and Japan: a brief introduction and comparison. Evid-Based Complement Altern Med 32, 546–555

Patriani Justo CM, dé Andrea Gomes MH (2008) Conceptions of health, illness and treatment of patients who use homeopathy in Santos, Brazil. Homeopathy 97(1):22–27

Reid R, Steel A, Wardle J, Trubody A, Adams J (2016) Complementary medicine use by the Australian population: a critical mixed studies systematic review of utilisation, perceptions and factors associated with use. BMC Complement Altern Med 16(1):176

Research America (2016) US investments in medical and health research and development. https://www.researchamerica.org/sites/default/files/2016US_Invest_R&D_report.pdf. Accessed 20 Apr 2018

Sharma U (1992) Complementary medicine today. Routledge, London

Shaw D (2010) Homeopathy is where the harm is: five unethical effects of funding unscientific 'remedies'. J Med Ethics 36(3):130–131

Shaw D (2011) Homeopathy and medical ethics. Focus Altern Complement Ther 16(1):17–21

Singh S, Ernst E (2008) Trick or treatment?: alternative medicine on trial. Bantam, London

Smith K, Ernst E, Colquhoun D, Sampson W (2016) Complementary & alternative medicine (CAM): ethical and policy issues. Bioethics 30(2):60–62

Stambolovic V (1996) Medical heresy—the view of a heretic. Soc Sci Med 43(5):601–604

Stein H (1990) American medicine as culture. Westview Press, Boulder CO

Tataryn DJ (2002) Paradigms of health and disease: a framework for classifying and understanding complementary and alternative medicine. J Altern Complement Med 8(6):877–892

Teixeira MZ (2009) Homeopathy: a preventive approach to medicine? Homeopatía: una praxis médica preventiva? 8(29):155–172

Tyreman S (2011) Values in complementary and alternative medicine. Med Health Care Philos 14(2):209–217

White P (2000) What can general practice learn from complementary medicine? Br J Gen Pract 50(459):821–823

Wieland LS, Manheimer E, Berman BM (2011) Development and classification of an operational definition of complementary and alternative medicine for the Cochrane collaboration. Altern Ther Health Med 17(2):50–59

Witt C, Schützler L (2012) The gap between results from sham-controlled trials and trials using other controls in acupuncture research—the influence of context. Complement Ther Med. https://doi.org/10.1016/j.ctim.2012.12.005

Wittgenstein L (1967) Philosophische Untersuchungen. Philosophical investigations. Translated by GEM Anscombe. Wiley-Blackwell, New Jersey

World Health Organization (2000) General guidelines for methodologies on research and evaluation of traditional medicine. World Health Organization, Hong Kong SAR

World Health Organization (2013) WHO traditional medicine strategy: 2014–2023. World Health Organization, Hong Kong SAR

Chapter 2
Defining Health

Abstract Unless we have a shared understanding of what health means, and what health looks like, we cannot agree upon what is needed to achieve it. However, the most commonly cited definition, that of the World Health Organization, has been subjected to a barrage of criticism because it is allegedly outdated and impossible to operationalise. More recent interpretations are aligned with concepts such as adaption, resilience and salutogenesis. As possibly the healthiest people in the world, the people of Okinawa in Japan have much to teach us about the qualities that are essential for health and wellbeing. Through examination of these illuminating topics, the meaning of health is described.

Keywords Health · Definition · World Health Organization · Adaption Resilience · Salutogenesis · Longevity · Okinawana

2.1 Setting Aims and Objectives in Health Care

Prior to undertaking an analysis of ethical issues associated with T&CM, it is necessary to explain what I mean by *health* as this, presumably, is the ultimate goal in health care. If we are to assume that health (or increased potential for health) is the main goal of medical interventions, then it follows that we need to have some agreement about what health means in order to set relevant objectives. If health is the ultimate aim of health care then a shared understanding of what health means is necessary to reach agreement about the strategies to be put in place, and services delivered, to help achieve it. As Fiona Godlee, editor of the *British Medical Journal,* remarked: '.... if health is the goal of healthcare and research, we need to know what it looks like and how to measure it' (Godlee 2011).

Those who work in healthcare might benefit from contemplating the approach of educationalists who appreciate the importance of clearly distinguishing between the identification of aims and the setting of objectives. In educational terms 'aims' are understood as statements of intent or aspiration; of what it is hoped to achieve. Objectives, on the other hand, are goals or steps that are set for meeting the aim; they are more specific and define measurable outcomes. Once the aim is identified, the objectives follow; to set objectives without a clear understanding of an aim makes no sense.

This distinction between aims and objectives, as widely utilised in education, is adopted here. This chapter is devoted to identification of a meaning of 'health' that will then be assumed, for the sake of argument in this analysis, as the main aim of health care. From this meaning of health, certain ethical implications for the provision of health care, and setting of objectives, are inferred. For this task, it is logical to begin with the examination of the most widely referenced definition of health; namely that of the World Health Organization (WHO).

2.2   The World Health Organization's Definition of Health

While there is a lack of broad agreement about how to define health, and a subsequent lack of a clear aim in health care, there have been numerous attempts to generate a definition of health (Van De Belt et al. 2010). Debate about the meaning of health and disease has been ongoing since antiquity and there have been many proposed definitions. One definition, however, stands out above all others in terms of breadth of acceptance and impact; the definition of health from the WHO, adopted by an international health conference in 1946, entered into force on 7th April 1948 and has not been altered since:

> Health is a state of complete physical, mental and social wellbeing and not merely the absence of disease or infirmity (WHO 1946).

Revolutionary in its time, this definition was intended to provide a transformative vision of 'health for all'; one that went beyond the prevailing negative conception of health based on an 'absence' of pathology (Larson 1996). However, it has been the subject of criticism since its inception as many have struggled to understand how its meaning could be operationalised. Furthermore, it is considered by some to be redundant because the WHO definition was, 'as much a political statement as a public health statement' (ibid.). In this post-war era, when there was acute awareness of the health status of whole societies, it was only logical that human health be placed in a broader social context. Brock Chisholm, the first Director General of the World Health Organization (1948–1953), and member of the group who formulated the WHO definition, describes a profound pessimism about humankind and society during that period writing:

> We have responsibility for social health, for being able to live in peace and contributing to the welfare of other people. The social responsibility of the individual has never been recognized before on such a wide international basis (Chisholm 1948).

Of the broad discourse surrounding the WHO definition, the four most significant objections are summarised here.

2.2.1 Utopian Nature

It has been argued that inclusion of the word 'complete' in the WHO definition implies that health is a state in which everything is perfect (Garner 1979). In other words, the definition is necessarily idealistic or utopian in nature. Anything less than complete absence of symptoms, and complete wellbeing on all levels, fails to achieve the requirements for health. As Bellieni and Buonocore (2009) assert, 'To define 'health' as 'complete wellbeing' would thus leave us with a 'null set' of persons as actually possessing health'.

Most people who claim to be healthy are aware of minor complaints from time to time and subjective feelings of wellbeing vary from day to day (Smith 2008). Additionally, it would not be considered 'healthy' for a person to maintain 'complete' mental wellbeing following a trauma such as loss of a loved one where grief would be considered a normal and healthy response. It is entirely conceivable that a person can be healthy without being in a state of 'complete physical, mental and social wellbeing'. As Callahan so eloquently points out, '...it is doubtful that there ever was, or ever could be, more than a transient state of 'complete physical, mental and social wellbeing' for individuals or societies; that's just not the way life is or could be' (Callahan 1973).

It could be argued that the WHO definition is aspirational and as such represents an ideal. However, even if we accept that the definition describes an aspirational goal, it is impossible to identify the steps needed to reach that goal, 'because 'complete' is neither operational nor measurable' (Huber et al. 2011). Or, as Doll puts it, 'This is a fine and inspiring concept and its pursuit guarantees health professionals unlimited opportunities for work in the future, but is not of much practical use' (Doll 1992).

2.2.2 Medicalisation of Health

Medical science is making such remarkable progress that soon none of us will be well.

Aldous Huxley

A further consequence of reference to 'complete' is that it is indirectly supportive of the increased medicalisation of health care (Huber et al. 2011). As new discoveries are made, such as new ways of diagnosing potential problems or new ways of treating problems, then the boundaries between health and ill-health are redrawn. Diseases are redefined, new drugs become available and ideas about what is healthy change accordingly: 'With the progress of medicine, individuals who are declared healthy today may be found to be diseased tomorrow' (Sartorius 2006).

For example, national guidelines in the UK about who should be prescribed one of the cholesterol lowering drugs, statins, have been subject to frequent changes. Initially statins were prescribed for people who had both high levels of serum cholesterol and a history of cardiovascular disease(CVD) but current guidelines (National Institute for Health and Care Excellence 2014) propose statins for anyone who has a 10%

or greater *risk* of CVD. Consequently, statins are the most widely prescribed drug treatment in the UK and yet the use of statins in low-risk patients without CVD remains a matter of intense debate. With thresholds for interventions constantly changing, an emphasis on 'complete physical wellbeing' could lead to ever increasing groups of people becoming eligible for treatment even when only few people might benefit, resulting in higher levels of medical dependency and risk (Huber et al. 2011). In this way, the WHO definition could even be counterproductive, leading to practice that is driven by developments in medicine rather than by what is most beneficial for the health of individuals.

2.2.3 Wellbeing and Health

The WHO definition equates health with a state of physical, mental and social *wellbeing,* but critics have argued that health and wellbeing are simply not the same thing. Saracci (1997), for example, asserts that the definition corresponds more closely to happiness than health, citing Freud in support of his argument who, after stopping smoking cigars for health reasons wrote:

> I learned that health was to be had at a certain cost.... Thus now I am better than I was, but not happier (Freud, cited by Saracci 1997).

Furthermore, the equation of health with wellbeing is of limited usefulness to healthcare professionals. As Fulder (1998) points out, 'there are no wellbeing medicines in the pharmacopoeia'; in other words, we cannot rely upon pharmaceuticals for our wellbeing. According to the Organisation for Economic Cooperation and Development (OECD), when we talk about human wellbeing we are referring to the quality of people's 'experience of life' and the quality of that experience is affected by many factors, health being one of them (OECD 2013). It is clear from both national and international studies that wellbeing is regarded as a complex issue with many contributing factors, including economic, social and environmental conditions as well as levels of health. Health matters for wellbeing, but it is possible to experience high levels of wellbeing even when apparently dealing with compromised health. For example, in their qualitative study Albrecht and Devlieger (1999) confirmed previous postulation of the so-called 'disability paradox'; they discovered that people with serious and persistent disabilities can report that they experience high levels of wellbeing even when, to most external observers, these individuals seem to live an undesirable daily existence.

2.2.4 The Changing Patterns of Disease

A further objection to the WHO definition of health is that the disease burden, particularly of high income nations, has altered significantly since the definition was introduced. In the first half of the twentieth century the majority of illness resulted from infectious disease and those who did have chronic diseases could expect an early death (Huber et al. 2011). During the Second World War there was a peak in infant mortality rates, mainly due to increased pneumonia and bronchitis (Macfarlane et al. 2000). While these may have been due to fuel shortages during the very cold winters of 1940 and 1941, and also to the disorganisation caused by evacuations early in the war, set against this backdrop it is easy to understand why a broad definition of health, inclusive of societal factors, would be welcomed.

However, aging with chronic illness has become the norm in the Western world, with an increasing number of people living with disabilities from physical complaints, such as musculoskeletal disorders, as well as dementia and Alzheimer's disease (Murray et al. 2013). This invariably raises concerns about the WHO definition of health, as it declares all those living with chronic disease or disability 'unhealthy'. With findings that people who live with disabilities can indeed regard themselves as healthy, there is a need for a 'conceptual disentangling' of health from disability (Drum et al. 2008). Changing patterns of demographics and disease demand a fresh look at the way in which health is defined, towards one in which people can aim for high levels of health and wellbeing in spite of physical or mental impairment. To do otherwise would be to label the majority (if not all) of the population as unhealthy.

In summary, while the WHO definition may have been appropriate and even revolutionary in its time, it is now of poor pragmatic value and consequently, it is of limited usefulness for either healthcare professionals or individuals. Given that the WHO definition is of limited value for consideration of what 'health' actually comprises, which is, of course, the crux of the matter in hand, it seems appropriate to ask the question, 'what does health look like'? To this end, I contemplate the attributes of certain populations, who are broadly accepted to be the healthiest communities in the world.

2.3 What Does Health Look Like?

At seventy you are but a child, at eighty you are merely a youth, and at ninety if the ancestors invite you to heaven, ask them to wait until you are 100…. And then you might consider it. An Okinawan proverb (Willcox et al. 2013).

Good health does not result exclusively from having access to the best healthcare systems. There are many factors involved, as is evident from examination of populations with apparently high levels of healthiness. In 2005, the National Geographic published an article about three distinct communities who have the greatest longevity in the world: Silanus, a mountainous region in Sardinia; a group of Seventh Day Adventists in Loma Linda, California; and Okinawa, an 800mile archipelago of one large and 160 tiny islands in Japan (Buettner 2005). Although each of these communities has become a focus of study because of the tendency towards great longevity of their peoples, researchers have found that they also exhibit much lower rates of chronic disease and more years of good life. Indeed, from studies around the world, it is widely accepted that, on average, the older people get, the healthier they have been in their lifetime (Hitt et al. 1999). The elderly of all three communities are remarkable in numerous ways but the apparent healthiness of elderly Okinawans is quite extraordinary.

In 1975, Dr. Makoto Suzuki began a population-based study of centenarians, and other selected elderly, in Okinawa, Japan, which is now known as the Okinawa Centenarian Study (OCS). When Suzuki first began his studies, he found an unusually large number of centenarians, many of whom were exceptionally healthy, with remarkably low rates of heart disease and cancer (Willcox et al. 2001). Okinawans boast the longest life expectancy in the world, but the OCS is not just about how many years have been lived: to be considered 'successfully aging' the individual should have enjoyed a high quality of life throughout life up to, and including, a 'super-elderly' stage (Suzuki et al. 2004). According to Willcox et al. (2013): 'they (Okinawans) have slim lithe bodies, sharp clear eyes, quick wits, passionate interests and the kind of Shangri-la glow of youthfulness we all covet'. Other signs of good health in elderly Okinawans include: impressively young, clean arteries; low levels of cholesterol; a low risk for hormone-dependent cancers; strong bones; sharp minds; slim bodies; natural menopause; healthy levels of sex hormones; low stress levels; low levels of depression and excellent pyschospiritual health (Willcox et al. 2007). Many dietary and lifestyle factors are believed to contribute to the good health of Okinawans, including low caloric intake, high vegetables/fruits consumption, higher intake of 'healthy' fats, lower intake of 'unhealthy' fats (from animal sources), high fibre intake, low body fat level, and high levels of physical activity (Willcox et al. 2009). However, there are also other factors at play.

The high levels of health experienced by Okinawans do not stem from stress-free lives or perfect living circumstances. On the contrary, most had bounced back from significant emotional ordeals and losses including poverty, war, oppression and other hardships, with amazing resilience. A seemingly crucial factor in this ability to 'bounce back' appears to be the high levels of social support, from strong social networks and close family ties, believed by many to increase protection against illness (Willcox et al. 2013). In fact, many studies have demonstrated that social ties can improve the strength and resilience of the immune system. Socially connected people are less prone to stress, while chronic stress is believed to wear down the body over time and is implicated in, amongst other things, an increased risk of heart disease, depression, and reduced immune function (Fagundes et al. 2012). Elderly Okinawans, in spite of their exposure to potentially stressful situations, exhibit low levels of negative emotions (Willcox et al. 2013), an attribute they share with other super-elderly. As Buettner described, 'After interviewing more than fifty centenarians on three continents, I've found every one likable; there hasn't been a grump in the bunch' (Buettner 2005).

One last, and potentially relevant, factor about the health of Okinawans is that they incorporate both Eastern and Western healing methods into their health system. Okinawa, Japan and Hong Kong, are the top three areas of the world for life expectancy and all incorporate both Eastern and Western medical approaches (Willcox et al. 2013).

While the characteristics of healthy Okinawans clearly illustrate a picture of what a healthy person may look like, it cannot be assumed that these attributes are necessary conditions for health. Indeed, these characteristics may be valuable indicators of health and a healthy lifestyle but whether or not they are all necessary for health is uncertain. Nevertheless, study of the Okinawans can aid with visualisation of the potential for health. If our aim is to be aspirational but realistic, and these are whom we consider as the healthiest people in the world, then our aim should be to help people achieve a state that is similar in some way to, but not necessarily the same as, that of the Okinawans.

2.4 Towards a New Definition of Health

In December 2008, two members of the Centre for Global eHealth Innovation at the University of Toronto, Alejandro Jadad and Laura O'Grady (2008), sparked new debate about how health should be defined when they called for broad input in an editorial of the British Medical Journal (BMJ). Jadad and O'Grady created a blog on the BMJ website, inviting anyone who had internet access to comment upon, challenge, or try to enhance the WHO definition of health. The efforts of Jadad and O'Grady motivated the Health Council of the Netherlands and the Netherlands Organisation for Health Research and Development, to support a two-day invitational conference entitled, '*Is health a state or an ability? Towards a dynamic concept of health.*' In December 2009, thirty-eight invited participants, representing a variety of stakeholders, came together to consider whether useful descriptions of health could be found for the perspectives of the different stakeholders. There was broad support at this meeting for moving from the 'static formulation' of the WHO definition towards a more dynamic definition. The outcome was a proposal for a new conceptualisation of health as:

> the ability to adapt and to self-manage' when facing physical, mental, and social challenges (Huber et al. 2011).

The proposed Huber et al. definition is suggestive of a marked shift away from a disease-focussed, or pathogenic, approach in healthcare towards a more positive and proactive stance. Broad acceptance of the pathogenic model in healthcare has resulted in the assumption that disease prevention, treatment and management are the best route to better health (Becker et al. 2010) and, at first glance, this definition appears to represent a radical change in thought and opinion. As Richard Smith, former editor of the BMJ, observed, the pathogenic approach in healthcare is pervasive:

> But what is health? For most doctors that's an uninteresting question. Doctors are interested in disease, not health. Medical textbooks are a massive catalogue of diseases. There are thousands of ways for the body and mind to go wrong, which is why disease is so interesting (Smith 2008).

While the move from the WHO definition to the Huber et al. definition might appear radical, the new definition was not conjured out of thin air. On the contrary, it is rooted in a growing change in perspective that has been developing over a number of years. Many health practitioners have described problems with a disease-oriented approach that is necessarily insufficient if health is not simply the mere absence of disease.

Simultaneously, there has been a marked increase in interest and discussion about the human capacity to respond, in a positive manner, in the face of adversity. Some individuals have the ability to achieve or retain a state of health and wellbeing even when faced with stresses of a physical, mental or social nature, as is markedly apparent in the peoples of Okinawa. Many theories about health have emerged that are, in certain ways, related to this observation, three of which have been the subject of significant interest and development. These are the concepts of salutogenesis, resilience and robustness.

2.4.1 Salutogenesis

The word 'salutogenesis' stems from the Latin term *salus* (health) and the Greek term *genesis* (origin) and was first introduced by sociologist Aaron Antonovsky in 1979. Antonovsky spent many years studying the ways in which people respond to stress, how they manage stress and how they stay well, and observed that some people achieve health in spite of exposure to potentially disabling stress factors (Antonovsky 1979). The salutogenic approach to health and wellbeing offers a stark contrast to the pathogenic model as it is focussed upon health and looks prospectively at how best to promote health and wellbeing. According to Antonovsky, health can be viewed as movement in a continuum on an axis between total ill health (disease) and total health (ease); movement in either direction can be affected by a number of factors. People are continually exposed to stress factors that encourage disease. The body is able to deal with most stresses without significant problems because of resources of both genetic and acquired character, as well as social and material in nature, such as understanding, intelligence, social support, family ties, culture, religion, and lifestyle factors (Antonovsky 1987). Stress factors will cause the coping mechanisms to fail whenever coping resources are not robust to deal with the current situation. Ultimately, it is the balance between stress factors and the coping resources that determines whether a factor will be pathogenic, neutral, or salutary.

2.4.2 Resilience

The word 'resilience' stems from the Latin term *resilire* (to jump back) and the concept was originally developed and explored in agricultural and ecological systems, where it is commonly used to refer to the capacity to resist shocks and/or the speed of recovery after some disturbance or pressure (Haimes 2009). The concept has also been widely adopted in other fields including sociology and, in particular, in psychology, where it is taken to mean the ability of people to cope positively with stressful situations (Döring et al. 2013). Psychological studies of resilience in particular groups of individuals who experience stress, such as military personnel (Sudom et al. 2014) and sports personnel (Sarkar and Fletcher 2014), are commonplace and the numbers of such studies are growing. Resilience is more than simply the ability to respond to stress; resilience is a coping mechanism that helps to keep systems (including the human body) in a healthy and functioning state.

2.4.3 Robustness

The word 'robustness' stems from the Latin term *robustus* (hard and strong) and also describes a coping mechanism. However, this concept was originally developed and employed in engineering and biology. Robustness has been described as a property that allows a system to maintain its functions despite external and internal perturbations (Sastry and Bodson 2011). It is an emergent property and cannot be understood by looking at the individual components of an organism. Complex biological systems must be robust in order to withstand environmental and genetic perturbations, and evolution often selects traits that might enhance the robustness of the organism. Hence, robustness is ubiquitous in living organisms (Kitano 2004). One example of how deficiencies in robustness have a direct effect on human health is the ability to cope with extreme weather conditions (Ivanovas et al. 2007). People who live permanently in more extreme weather environments seem to develop a much greater ability to cope with the extremes than people who only experience extreme weather environments on a temporary basis.

The proposed Huber et al. definition of health is rooted in these three overlapping concepts that are receiving increased levels of interest around the world. Like the Huber et al. definition, they all point towards there being a dynamic process that is stimulated when an individual is under stress, and which serves to preserve or promote health through adaption. However, for any new definition to be broadly accepted, it needs more than wide scale interest; it needs to demonstrate that it has real world relevance and can withstand the critique that is so heavily levied at the WHO definition of health.

2.5 The Implications of the Huber et al. Definition of Health

As longevity and numbers of people living with multi-morbidity and disability increase, so the need for change in the way we conceive of health becomes more pressing. In contrast to the WHO definition, according to this new definition, it is entirely possible for people with disabilities and chronic complaints to be considered as healthy if they have adapted and are coping well with their circumstances. However, for the definition to be operational we would need to develop ways in which to assess and measure people's ability or potential to adapt so that there are means of assessing whether interventions are helpful.

Huber believes that the formulation 'health as the ability to adapt' could help to make health effects easier to measure and assess and, similarly, Antonovsky believed that salutogenesis should be the theoretical basis for developing, testing and implementing plans and practices that enhance health and wellbeing. To this end, he developed tools that have been tested in practice, primarily in Scandinavian countries (Eriksson and Lindström 2007) but the scales involved have not received broad based testing globally. Furthermore, we need a great deal more understanding about how to promote salutogenesis, resilience or robustness. Acceptance of this definition will mean a new way of thinking in healthcare, new ways of treating people, new ways of assessing health and new ways of establishing priorities.

The Huber et al. definition of health is much more in keeping with T&CM philosophies and conceptions of health than the WHO definition. There are currently some radical differences in approach between T&CM and conventional medicine, at least in part due to the differences in conceptualisation of health. For example, some symptoms that a T&CM practitioner might view as signs of a *healthy adaptive response* for most people, such as fever, nasal discharges or diarrhoea, might be viewed as *pathology* by a conventional medic. Adoption of the new definition would require a shift in thinking towards one that appreciates physiological signs of adaption, such as low-grade fever during infection, as a sign of health rather than disease. If both conventional and T&CM practitioners agree upon how to conceptualise health, this might enable them to work towards common goals.

From studies of the healthiest people in the world we have indications of factors that may help to improve adaption and coping. Apart from genetic factors, there are important lifestyle aspects such as maintaining activity, feeling socially connected and having a good diet. A positive outlook also appears to be of vital importance, and given the enormous global burden of psychiatric disease, this is an issue that will need broad attention. Furthermore, acceptance of health as the ability to adapt would necessitate a revision of the way in which decisions are made about healthcare priorities, policy and funding, to account for the potential effects on coping and adaption of specific treatments and interventions. Interestingly, 'access to the highest quality healthcare services' is not in the list of factors attributed to enhancing the health of the super-elderly, implying that the goal of achieving higher levels of health may not entail additional burden on an increasingly overstretched healthcare service.

In conclusion, the definition of health as *'the ability to adapt and to self-manage' when facing physical, mental, and social challenges'* (Huber et al. 2011) is in keeping with current trends across a broad domain of disciplines, including conventional medicine and T&CM. It describes an aspirational aim and has much greater operational value that the WHO definition. When held as the stated aim, it has certain implications for objectives in health care practice that will be taken into account when analysing ethical issues in T&CM.

References

Albrecht GL, Devlieger PJ (1999) The disability paradox: high quality of life against all odds. Soc Sci Med 48(8):977–988

Antonovsky A (1979) Health, stress, and coping. Jossey-Bass, London

Antonovsky A (1987) Unraveling the mystery of health: How people manage stress and stay well. Jossey-Bass, London

Becker CM, Glascoff MA, Felts WM (2010) Salutogenesis 30 years later: Where do we go from here? Int Electron J Health Educ 13:25–32

Bellieni CV, Buonocore G (2009) Pleasing desires or pleasing wishes? A new approach to health definition. Ethics Med Int J Bioeth 25(1):7–10

Buettner D (2005) Who's best at living longest: the secrets of longevity. The National Geographic, November 2005

Callahan D (1973) The WHO definition of 'health'. Hast Cent Stud 1(3):77–87

Chisholm B (1948) Organization for world health. Ment Hyg 32:364–371

Doll R (1992) Health and environment in the 1990s. Am J Public Health 82(7):933–941

Döring TF, Vieweger A, Pautasso M, Vaarst M, Finckh MR, Wolfe MS (2013) Resilience as a universal criterion of health. J Sci Food Agric 95(3):455–465

Drum CE, Horner-Johnson W, Krahn GL (2008) Self-rated health and healthy days: examining the "disability paradox". Disabil Health J 1(2):71–78

Eriksson M, Lindström B (2007) Antonovsky's sense of coherence scale and its relation with quality of life: a systematic review. J Epidemiol Community Health 61(11):938–944

Fagundes CP, Gillie BL, Derry HM, Bennett JM, Kiecolt-Glaser JK (2012) Resilience and immune function in older adults. Annu Rev Gerontol Geriatr 32:29–47

Fulder S (1998) The basic concepts of alternative medicine and their impact on our views of health. J Altern Complement Med 4(2):147–158

Garner L (1979) The NHS: your money or your life. Penguin Books, New York

Godlee F (2011) What is health? BMJ 343(4817):1

Haimes YY (2009) On the definition of resilience in systems. Risk Anal Int J 29(4):498–501

Hitt R, Young-Xu Y, Silver M, Perls T (1999) Centenarians: the older you get, the healthier you have been. Lancet 354(9179):652

Huber M, Knottnerus JA, Green L, van der Horst H, Jadad AR, Kromhout D, Leonard B, Lorig K, Loureiro MI, van der Meer JWM, Schnabel P, Smith R, van Weel C, Smid H (2011) How should we define health? BMJ 343:d4163

Ivanovas G, Tomaras V, Papadioti V, Paritsis N (2007) Human robustness and conscious purpose in contemporary medicine. Kybernetes 36(7/8):972–984

Jadad AR, O'Grady L (2008) How should health be defined? Br Med J 37:a2900

Kitano H (2004) Biological robustness. Nat Rev Genet 5(11):826–837

Larson JS (1996) The World Health Organization's definition of health: social versus spiritual health. Soc Indic Res 38(2):181–192

Macfarlane A, Mugford M, Henderson JE, Furtado A, Stevenson JS, Dunn A (2000) Birth counts: statistics of pregnancy and childbirth. Stationery Office, London

Murray CJL, Richards MA, Newton JN, Fenton KA, Anderson HR, Atkinson C, Bennett D, Bernabé E, Blencowe H, Bourne R, Braithwaite T, Brayne C, Bruce NG, Brugha TS, Burney P, Dherani M, Dolk H, Edmond K, Ezzati M, Flaxman AD, Fleming TD, Freedman G, Gunnell D, Hay RJ, Hutchings SJ, Ohno SL, Lozano R, Lyons RA, Marcenes W, Naghavi M, Newton CR, Pearce N, Pope D, Rushton L, Salomon JA, Shibuya K, Vos T, Wang H, Williams HC, Woolf AD, Lopez AD, Davis A (2013) UK health performance: findings of the Global Burden of Disease Study 2010. Lancet 381(9871):997–1020

National Institute for Health and Care Excellence (2014) Lipid modification: Cardiovascular risk and the modification of blood lipids for the primary and secondary prevention of cardiovascular disease. National Health Service, UK

Organisation for Economic Cooperation and Development (2013) Measuring well-being for development. Global Forum on Development, Paris, 4–5 April 2013

Saracci R (1997) The World Health Organisation needs to reconsider its definition of health. BMJ 314(7091):1409–1410

Sarkar M, Fletcher D (2014) Psychological resilience in sport performers: a review of stressors and protective factors. J Sports Sci 32(15):1419–1434

Sartorius N (2006) The meanings of health and its promotion. Croat Med J 47(4):662–664

Sastry S, Bodson M (2011) Adaptive control: stability, convergence and robustness. Courier Corporation, Nashville

Smith R (2008) The end of disease and the beginning of health. BMJ Group (Blog), 10 December 2008

Sudom KA, Lee JEC, Zamorski MA (2014) A longitudinal pilot study of resilience in Canadian military personnel. Stress Health J Int Soc Investig Stress 30(5):377–385

Suzuki M, Willcox B, Willcox C (2004) Successful aging: secrets of Okinawan longevity. Geriatr Gerontol Int 4:S180–S181

Van De Belt TH, Engelen LJLPG, Berben SAA, Schoonhoven L (2010) Definition of Health 2.0 and Medicine 2.0: a systematic review. J Med Internet Resc 12(2):e18

Willcox BJ, Willcox DC, Suzuki M (2001) Evidence-based extreme longevity: the case of Okinawa, Japan. J Am Geriatr Soc 49(4):S135–S136

Willcox DC, Willcox BJ, Shimajiri S, Kurechi S, Suzuki M (2007) Aging gracefully: a retrospective analysis of functional status in Okinawan centenarians. Am J Geriatr Psychiatry 15(3):252–256

Willcox DC, Willcox BJ, Todoriki H, Suzuki M (2009) The Okinawan diet: health implications of a low-calorie, nutrient-dense, antioxidant-rich dietary pattern low in glycemic load. J Am Coll Nutr 28(Suppl):500S–516S

Willcox BJ, Willcox DC, Suzuki M (2013) The Okinawa way. Penguin Books, New York

World Health Organiszation (1946) Preamble to the Constitution of the World Health Organization as adopted by the International Health Conference. World Health Organisation, New York

Chapter 3
An Ethical Matrix for Traditional and Complementary Medicine

Abstract Ethical challenges to the use of T&CM are wide-ranging. The vast majority of published concerns are associated with the potential for harm to human users but there are also concerns for the wellbeing of animals and the environment. The ethical matrix is a bioethical methodology that has proven value across a number of disciplines to aid identification and analysis of ethical issues. Its use entails the mapping of ethical challenges against the applied ethical principles of wellbeing, autonomy and justice. Using the ethical matrix, ethical challenges related to T&CM are identified. Following completion of the matrix, the full extent of ethical issues relating to humans, animals and the environment becomes evident.

Keywords Traditional medicine · Complementary medicine
Alternative medicine · Ethical matrix · Principlism · Wellbeing · Autonomy
Justice

3.1 The Ethical Matrix

Analysis of ethical challenges to T&CM could be undertaken in many different ways but the methodology of choice is that of the ethical matrix, as developed by Ben Mepham (1995). The ethical matrix is a bioethical methodology that was originally designed to help evaluate public policy decisions involving agriculture and food production. However, since its inception, this method has been applied widely in a number of fields and has proven to be a versatile tool for analysing ethical issues. The principal aim of the matrix is to assist, 'rational decision making by articulating the ethical dimensions of any issue in a manner which is transparent and broadly comprehensible' (Mepham 2005a).

One advantage of using this method is that the matrix sets out a framework to help identify a broad range of ethical issues that might not otherwise be obvious. It does so by creating a formal structure for identification of the parties (stakeholders) involved in a given situation and their potentially conflicting interests (Fig. 3.1).

Respect for:	Wellbeing	Autonomy	Justice
Stakeholders			
A			
B			
C			

Fig. 3.1 The ethical matrix

Identified stakeholders are listed on the vertical axis and the interests of each considered with respect to the three ethical principles set out on the horizontal axis.

3.2   The Conceptual Basis of the Ethical Matrix

The field of ethics is commonly divided into three broad areas: meta-ethics, normative ethics and applied ethics.[1] Meta-ethics concerns the status, foundations, and scope of moral values; it questions the nature, and indeed the very existence of morality. Normative ethics is concerned with the standards and principles used to determine whether something is right or good and is itself commonly divided into various sub-branches such as, consequentialist theories, deontological theories, and virtue-based theories. Applied ethics, is concerned with the moral permissibility of specific actions and practices. While these three areas of ethics may appear to be distinct they are also interrelated. For example, the use of an applied ethics approach often draws upon certain normative ethical theories such as consequentialism or deontology.

Applied ethical analysis is often conducted through the use of an 'ethical framework' because frameworks can provide a consistent and structured approach to the analysis. An extremely large number of ethical frameworks have been developed. Some are aligned with particular normative ethical theories, such as: the consequentialist framework, the duty framework, and the virtue framework (Bonde and Firenze 2013). The ethical framework selected for this enquiry, the ethical matrix, is an example of a cross-normative theory framework that can be adapted for use in different circumstances and designed to reflect the concerns of different stakeholders.

The three ethical principles used in the ethical matrix are drawn from the prominent principlist approach in biomedical ethics, introduced by Tom Beauchamp and James Childress in 1979, who stated that the following four *prima facie* principles lie at the core of moral reasoning in health care: respect for autonomy, beneficence, non-maleficence and justice. In the opinion of Beauchamp and Childress, these four principles are part of a 'common morality'; an approach that 'takes its basic premises

---

[1] There are other ways of categorising the different dimensions of ethics but this is the most commonly described.

directly from the morality shared by the members of society – that is, unphilosophical common sense and tradition' (Beauchamp and Childress 1994). In the construction of the ethical matrix, Mepham combines the two principles of beneficence and non-maleficence into one principle, that of 'wellbeing'.

The principlist approach is derived from normative ethical theories, but it is not aligned to any one single theory. While Beauchamp and Childress (2001) claim that these principles are commonly understood and accepted within society, and thus have a broad degree of support, they also assert that they are drawn from two normative ethical traditions: the duty-based moral philosophy (deontological) of Immanuel Kant and the outcome-based (consequentialist) ethics of Jeremy Bentham and John Stuart Mill. Mepham, in turn, extends the conceptual basis of his approach stating that the three principles in the ethical matrix represent *three* dominant perspectives in normative ethics, namely: utilitarianism (wellbeing), deontology (autonomy), and Rawlsian social contract theory (justice) (Mepham 2005b). The ethical matrix has been praised for its cross-ethical-theory approach, thus avoiding critique that might be levied at a single normative ethical; the cross-theory approach facilitates analysis from a variety of perspectives (Cotton 2009). However, the design of the ethical matrix also has its limitations and these need to be explicitly acknowledged.

The ethical matrix has been criticised for the lack of suitable deliberative mechanisms for enabling the ethical decision making needed to assist policy development (ibid). As stressed by Schroeder and Palmer (2003), 'The ethical matrix is helpful for fact finding in ethical debates but much less helpful in weighing the different ethical problems that it uncovers'.

Even when faced with a completed matrix, containing all the relevant information and representation of different stakeholders, a moral judgement must be exercised. Mepham, himself, acknowledges that circumstances will frequently arise where there are conflicts between different principles and where compromises will have to be made. In these situations, ethical evaluation or judgement requires a weighing or ranking of the different impacts (Mepham 2005a). In other words, ethical decision making is reliant upon the competency of the users' moral judgement, provoking the question of *who should be the judge*.

When the ethical matrix was first applied as a participatory tool it was used with a multi-stakeholder group involved in the discussion and the decision making process, enabling a 'bottom up' approach to ethical analysis (Kaiser and Forsberg 2001). However, the bottom-up approach is fraught with difficulties, such as who can be considered as representative of each of the stakeholder groups, who is capable of exploring political and economic implications, and how we can best represent the non-human stakeholder groups such as animals and the environment. The issue of which is best, the 'top down' or 'bottom up' approach, in application of the matrix is a subject of debate but Mepham believes the matrix can be usefully employed in both ways, depending upon the desired outcome (Mepham 2006). As Jensen et al. (2011) highlighted, successful use of the ethical matrix is dependent upon the users being able to 'put themselves in the shoes of others' rather than being reliant upon representatives of the stakeholder groups.

In this analysis a top-down approach to the use of the ethical matrix is inescapable. It involves the placing of oneself in the 'shoes of others' as far as possible in order to appreciate the ethical issues from the perspectives of different stakeholders. It also involves the weighting of certain principles where conflicts arise, for example between wellbeing and autonomy or wellbeing and justice.

A further considerable challenge for the ethical matrix arises directly from the adoption of a principlist approach, leaving the very essence of the ethical matrix open to the same extensive critique that principlism has received (Schroeder and Palmer 2003). While the controversy surrounding the merits of principlism is acknowledged, it is beyond the scope of this analysis to consider the pros and cons of principlism per se. There are undoubtedly limitations to a principlist approach to ethical analysis, but it has time-tested value, and continues to be the most dominant approach to ethical analysis in healthcare. Indeed, the book *Principles of Biomedical Ethics* by Beauchamp and Childress is the most influential book in modern bioethics and hence adoption of this approach should have broad-based appeal and comprehensibility.

3.3   The Ethical Matrix in This Analysis

There were three steps involved in formulating the ethical matrix for T&CM:

(1) Identifying and categorising published ethical challenges to T&CM through a broad literature search;
(2) Formulating the structure of the ethical matrix, including the selection of stakeholders; and
(3) Mapping the identified challenges onto the ethical matrix to reveal existing areas of debate and tension.

3.3.1   Identification of the Ethical Challenges to T&CM

In order to capture the full extent of existing debate about ethical challenges for T&CM, a systematic literature search of published academic papers was undertaken. As well as using general terms in my search, such as 'complementary', 'alternative' and 'traditional', I also searched for ethical concerns related to five specific forms of T&CM: acupuncture, homeopathy, herbal medicine, osteopathy and chiropractic as representing some of the most commonly used types of T&CM in the Western world. It was anticipated that this search would reveal the majority of previously identified ethical issues and ensure that the contents of the matrix are representative of a broad base of opinions and not simply my own.

Numerous papers mention potential ethical issues about healthcare practice but only those that described challenges to T&CM were reviewed in detail. Unfounded assertions (such as: 'homeopaths kill people') were excluded from the findings;

genuine ethical challenges were tabulated. Many repetitions of particular issues were discovered as time progressed and hence detailed records were only kept of the papers in which the issues were most clearly articulated, or that added some new perspective. Results of the literature search were organised and tabulated with description of the challenges, type of T&CM (where specified) and related stakeholder to be used in step 3.

3.3.2 Formulating the Structure of the Ethical Matrix: Choice of Stakeholders

My ethical matrix for the analysis of ethical challenges to T&CM adopts the same three principles originally suggested by Mepham (wellbeing, autonomy and justice) but the stakeholders need to be specific to the subject in hand. The particular stakeholders to be included in a matrix depends upon the issue under consideration, but the rule of thumb is that views represented in society should be reflected in the ethical matrix. Of primary importance is that those included should possess 'ethical standing'. In other words, they are subjects of ethical consideration in their own right, and not just means to others' ends (Mepham et al. 2006). Another factor of importance is that the range of stakeholders is broad enough to be inclusive of all legitimate concerns. In practice this can be challenging. As the number of stakeholders are increased, the matrix can become unwieldy and overcomplicated and hence, decisions need to be made about which groups possess ethical standing, and which to include to ensure that most, if not all, concerns are captured in the matrix.

The stakeholders included in the matrix for this analysis are:

Humans, as they are the providers and primary users of T&CM.
Animals, as they may also be recipients of T&CM in a domesticated or farming environment. In addition, many animals and animal products are used in the production of traditional medicines and they may also be used for research purposes during development and testing.
The environment, as the production, transport and storage of medicines has a significant environmental impact.

These three stakeholder groups were selected primarily because they were the most often mentioned in the published literature and hence they most clearly reflect the ethical challenges revealed by the literature. There are undoubtedly other potential stakeholders and some others were considered. For example, (Schroeder and Palmer 2003) recommend that 'future generations' should be included as a default stakeholder group on all occasions because they cannot intervene in the decision making process and yet are deeply affected by the outcome. However, in this analysis the considerations of future generations are no different to those of the 'humans' group in regard to wellbeing and autonomy and would differ only in terms of justice. For future generations this would involve the right to have at least the same access to, and choice of, T&CMs as the current generation of humans and this in turn is largely

Respect for:	Wellbeing	Autonomy	Justice
Humans			
Animals			
Environment			

Fig. 3.2 The structure of the ethical matrix in this analysis

dependent upon the sustainability of T&CM products and practices. As sustainability is also a major factor for consideration for the environment, I took the decision not to include future generations as a stakeholder group in their own right. The interests of future generations are, nonetheless, represented indirectly through consideration of humans and the environment.

Hence, the matrix in this enquiry takes the form shown below in Fig. 3.2.

3.3.3   Mapping of the Identified Challenges onto the Ethical Matrix

From the tabulated data produced in step 1, each identified ethical objection was extracted and positioned on the ethical matrix with respect to the relevant stakeholder and ethical principle. This resulted in a complex matrix with the vast majority of identified ethical challenges relating to humans, particularly the wellbeing of humans, as shown in Table 3.1.

Where the identified ethical challenges have bearing upon more than one of the ethical principles, a decision was made as to the most relevant position on the matrix so that the same challenges are not replicated in numerous cells. For example, the challenge 'Information on T&CM is unreliable and dangerously misleading' could be considered a threat to wellbeing, but it also has a direct bearing upon autonomy because people need reliable information for informed decision making. Hence, this challenge has been situated within the autonomy cell.

From this mapping exercise, it is clear that there are a number of different ethical issues to be considered for each stakeholder group. While a very large number of ethical challenges relating to the wellbeing of humans have been described, most cells of the matrix detail a smaller number of challenges. It cannot be assumed from this that there are fewer ethical issues relating to the other groups or concerns, but rather this is simply a reflection of what was discovered in the literature search. The main benefit of the mapping exercise is that it clearly reveals the primary ethical challenges and areas of conflict to be considered.

A note about the ascription of ethical challenges to animals and the environment in the matrix

Table 3.1 Mapping ethical challenges for humans onto the ethical matrix

Respect for wellbeing	Respect for autonomy	Respect for justice
Safety issues • T&CM has serious adverse effects **Adverse drug reactions** From herbal medicine • Toxicity • Drug/herb interactions • Allergies • Contamination • Interfere with results from lab tests Herbs and supplements are often regulated as food rather than medicine so they lack the mandatory efficacy and safety assurances required of pharmaceuticals From homeopathy • Contaminated medications • Allergic reactions • Poisonings **Adverse events** • Acupuncture has caused death and serious complications through infection and trauma • Chiropractic and osteopathy frequently result in mild adverse events and can cause vertebral arterial dissection • Rejection of, or delay in access, to effective conventional care can lead to serious consequences • T&CM practitioners are denying their patients more effective conventional treatment • Homeopaths and chiropractors frequently advise against immunisation • Claims about disease prevention are not based on science • T&CM providers interfere with doctors' prescriptions **Clinical competence and regulation** • There are concerns about clinical competence of practitioners because of variable standards in education and training • T&CM providers are often not medically trained • T&CM practitioners are not well regulated • Practitioners are often working alone, without governance measures in place **T&CM is anti-science** • An attitude of anti-science is promoted leading to a general weakening of support for science-based medicine • Government registration of regulatory bodies is tacit approval of T&CM • Philosophies of T&CM are not based on scientific principles • Any health service support for homeopathy could weaken patient confidence **Homeopathy is a threat to other T&CMs** • Funding homeopathy distracts from other more effective forms of T&CM • Homeopathy promotes a weakening of support for genuine T&CMs	**Patient choice, education and informed consent** • Many T&CM practitioners fail to gain formal consent from their patients • There are concerns about consent when parents substitute T&CMs for conventional treatments for their children • Consent cannot be determined because the risks are unknown • Effects are often not known or researched and so may violate utility in unknown ways • Conventional medical staff's lack of knowledge and lack of support creates a significant barrier to accessing information about T&CM • There is deliberate deception of the patients who are misled into thinking that treatments have an effect • Information about T&CM is dangerously misleading • Information about T&CMs is unreliable and incomplete • Claims about disease prevention are not based on science and are therefore misleading • Much of the research in T&CM is methodologically weak • There is a general lack of systematic reporting systems for adverse events • There is a widespread belief that T&CMs are safe because they are natural	**Affordability and access** • There is inequality and an unequal distribution of access to T&CM • In Europe, T&CM is primarily used by educated citizens of working age and with an above average income • Allocation of funding: research funds are scarce and should go to those areas where reasonably good evidence already exists • Funding should not be allocated to research of implausible treatments as this constitutes a waste of medical resources • Use of T&CM may result in double costs: T&CM may constitute additional expense over and above other healthcare costs • Use of T&CM can be unnecessary and costly • Cost-effectiveness has not been demonstrated • Health services should not fund treatments that have no evidence base

In his analysis of ethical concerns for animals in biotechnology, Mepham acknowledges that the ethical principles normally used in medical ethics need to be appropri-
ately translated to represent the interests of non-human stakeholder groups. Accordingly, for animals in biotechnology, Mepham interprets respect for wellbeing, autonomy, and justice as respect for welfare (freedom from pain and stress), freedom of
behavioral expression and respect for telos,[2] respectively (Mepham 2000). Similarly,
Mepham interprets the ethical principles of respect for wellbeing, autonomy and justice for applicability to the environment. In the case of the environment, respect for
wellbeing requires protection of the biota[3]; respect for autonomy demands maintenance of biodiversity and respect for justice calls for sustainability of biotic populations (Mepham 1995). However, there is no satisfactory philosophical rationale for
these interpretations. While considerations of animal dignity and behavioural freedom, as well as environmental biodiversity and sustainability, are extremely important, in this analysis these considerations are not ascribed to the particular principles
of autonomy and justice. Rather, it is assumed that, for the case of animals and the
environment, a strictly consequentialist approach is sufficient to reveal the major
ethical challenges and enable robust analysis. After all, it is possible to consider con-
straints upon autonomy and violations of the principle of justice under the principle
of wellbeing through analysis of the potential for harms and benefits.

This approach avoids the many complications that arise from debate concerning
the application of autonomy and justice to animals and the environment, and is in
keeping with an approach suggested by Häyry who found it to be more beneficial than
Mepham's approach in the case of bovine growth hormone (Häyry 2000). Hence, all
the ethical challenges related to the use of T&CM from animals and the environment
are included under respect for wellbeing in Table 3.2.

Mapping of the ethical challenges onto the ethical matrix has clearly revealed that
ethical considerations about the production and usage of T&CM are not limited to
the interests of human users. The interests of human users need to be considered
within this broader framework, and balanced against impact upon the environment
or animals. What follows is a brief summary of the main concerns.

Humans

Wellbeing
Concerns about safety can be broadly divided into adverse drug reactions and adverse
events. Adverse drug reactions are harmful side effects from use of a T&CM product
such as allergic reactions, toxicity, or overdose from a herbal medication. Adverse
events are normally associated with practitioner errors, negligence or incompetence
in application and include cases of trauma and injuries from acupuncture or chiropractic.

Additionally, and by far the most commonly cited adverse event associated with
use of T&CM, is that rejection of, or delay in access to, conventional care can

---

[2]Telos can be thought of as the inherent purpose of an entity or 'what it was made for'. Respect for
the telos of a lion, for example, would entail respect for its right to hunt, eat, sleep, reproduce, laze
in the sunshine etc. In other words, to fulfil it's potential as a lion.

[3]The animal and plant life in a given region.

Table 3.2 Ethical challenges to wellbeing for animals and the environment

	Respect for Wellbeing
Animals	**Harm in the production of products** • Many T&CM products are derived from animals • In China, more than 1500 animals are used in medicine and in India, 15–20% of Ayurvedic remedies are based on animal products • In Brazil, at least 354 animal species are used in medicinal products of which 21% are on one or more lists of endangered species • Many animals suffer during farming processes, extraction of medicinal products and/or slaughter • Use of T&CM encourages unethical animal killing for products of unproven value • Excessive or uncontrolled hunting to secure animal products for use in T&CM has led to the extermination of some species • The illegal trade of animal products for medicinal use has contributed to a decrease in animal populations **Harm in the testing of products** • Animals are routinely used in exploratory studies, often exposed to stress, pain, artificially induced diseases and ultimately killed
Environment	**Effects upon individual species** • A growing demand for standardised herbal products is putting pressure on selected high demand species • Many plants and animals used for medicinal purposes are becoming extinct as a result of demand for T&CM products • Reliance upon indigenous medicines and the resulting large scale commercial exploitation for urban markets are seen as a threat to biodiversity • Harvesting without planting, deforestation and the increased marketing of medicinal plants have resulted in the decline and sometimes near-extinction of several valued medicinal plant species around the world **Other effects on ecosystems** • A demand-oriented market and poor quality checks in T&CM has motivated producers to go for mass production without taking into account the finer details of plant cultivation • Most medicinal plants are collected from the wild in an uncontrolled manner and cultivated plants are often considered inferior **Limited resources** • Growing and harvesting, processing and packaging and transport of herbs is energy intensive • In turn, climate change may lead to the extinction of species, reduction in availability, and reduction in quality • In the future T&CMs may become even more expensive than conventional medicine if its resources are not readily available

lead to serious consequences; the underlying assumption being that conventional care is more effective than T&CM. In addition, it is assumed that practitioners of T&CM may fail to recognise serious conditions and/or interfere with a conventional treatment approach through contradictory advice about conventional medications and immunisation.

Further concerns about T&CM for human users arise from the wide variation in standards of education, training and regulation leading to questions about the clinical competence of practitioners. In addition, harm may occur in unpredicted ways because many forms of T&CM are poorly researched for efficacy or safety.

Autonomy

Concerns for autonomy are particularly related to the challenge that information is either simply not available or that it is deliberately misleading. Many people assume that T&CM products and practices are safe because they are natural and many practitioners fail to gain informed consent from their patients. There were particular concerns expressed for children whose parents may elect for T&CM treatments in preference to conventional treatments.

Justice

In many countries people pay out of pocket for T&CM treatments, leading to inequality of access. The use of T&CM can be costly and information about cost-effectiveness is scarce. Because the research evidence base for T&CM treatments is relatively small, any funding of such treatments is challenged on the grounds that funding should go into treatments where reasonably good evidence already exists. Funding for research into T&CM treatments is also challenged on the same grounds.

Animals

Wellbeing

Many forms of T&CM are used in veterinary practice and hence carry the same potential for risks as for human users, such as the risk of infection or trauma from acupuncture or the risk of toxicity from herbal medications. However, animals are clearly harmed in the manufacture of some T&CM products as a significant number are derived from whole or parts of animals. As well as the issue of whether it is ethically acceptable to make medicinal products for humans from animals, there are also concerns about how animals are treated during farming and slaughter. Restriction of the behavioural freedom of animals may cause suffering in the farming process, and pain may be inflicted during the extraction of T&CM products. The decrease in certain animal populations and extinction of some species from over-hunting for the use in T&CM products is especially concerning. When animals are used for research purposes in the production and testing of T&CM products, they can be exposed to stress, pain or artificially induced diseases.

Environment

Wellbeing

The production of herbal medicines, in particular, has a direct impact upon the environment as some plant species are in great demand. The growing and harvesting, processing and packaging, and transport of herbs can be energy intensive. Additionally, the carbon footprint associated with production and transport contributes to climate change that in turn affects species' viability. As a result, many forms of T&CM now used may not be available in the future and others may become scarce and more costly.

The high demand for some products can motivate producers to go for mass production without regard for consequences for the environment. Many plants are harvested in an uncontrolled manner and many plants and animals that are used for medicinal purposes are becoming extinct. On the other hand, many plants are awarded value because of their role in T&CM and direct local use can contribute to the preservation of some species and habitats. This needs to be weighed against the harm caused by disappearance of species (animals and plants) as a result of demand for T&CM products.

3.4   The Next Steps

It is clear, from my development of the ethical matrix for T&CM, that most published concerns are related to human users, but the inclusion of other stakeholders has revealed competing interests. For example, there are conflicts between humans and the animals that are used in the production of T&CM products and between humans and the environment that is being harmed through the carbon footprint and reduction in biodiversity as a direct consequence of T&CM use. The aim of this analytical process is to use the ethical matrix as a conceptual tool to arrive at recommendations for public policy decisions as intended by Mepham (2009) and, as such, the next step is to undertake in-depth analysis of the most ethically challenging features of T&CM as revealed by the matrix. Through this process, recommendations about the use of T&CM can be formulated on the basis of ethical decisions that prioritise the conception of health previously identified.

Given that the number of identified ethical challenges related to humans is vast, the analysis in this book is restricted to issues related to safety, considering the potential for both direct adverse effects (adverse drug reactions) and indirect adverse events. This is followed by an analysis of issues related to animals and the environment.

References

Beauchamp TL, Childress JF (1994) Principles of medical ethics. Oxford University Press, New York
Beauchamp TL, Childress JF (2001) Principles of biomedical ethics, 5th edn. Oxford University Press, New York
Bonde S, Firenze P (2013) A framework for making ethical decisions. Program in science and technology studies. https://www.brown.edu/academics/science-and-technology-studies/framework-making-ethical-decisions. Accessed 10 May 2018

Cotton M (2009) Evaluating the 'ethical matrix' as a radioactive waste management deliberative decision-support tool. Environ Values 18(2):153–176

Häyry M (2000) How to apply ethical principles to the biotechnological production of food—the case of bovine growth hormone. J Agric Environ Ethics 12(2):177–184

Jensen KK, Forsberg E-M, Gamborg C, Millar K, Sandøe P (2011) Facilitating ethical reflection among scientists using the ethical matrix. Sci Eng Ethics 17(3):425–445

Kaiser M, Forsberg E-M (2001) Assessing fisheries—using an ethical matrix in a participatory process. J Agric Environ Ethics 14(2):191–200

Mepham B (1995) An ethical matrix for animal production. New Farmer Grow 46:14

Mepham B (2000) A framework for the ethical analysis of novel foods: the ethical matrix. J Agric Environ Ethics 12(2):165–176

Mepham B (2005a) The ethical matrix: a framework for teaching ethics to bioscience students. Animal bioethics: principles and teaching methods. Wageningen Academic Publishers, Wageningen

Mepham B (2005b) A framework for ethical analysis. Bioethics: an introduction for the biosciences. Oxford University Press, Oxford

Mepham B (2006) The ethical matrix as a decision-making tool, with specific reference to animal sentience. In: Turner J, D'Silva J (eds) Animals, ethics and trade. Earthscan, London, pp 134–146

Mepham B (2009) The ethical matrix as a tool in policy interventions: the obesity crisis. In: Ingensiep HW, Meinhardt M (eds) Food ethics: utopia and reality. Springer, New York

Mepham B, Kaiser M, Thorstensen E, Tomkins S, Millar K (2006) Ethical matrix manual. Agricultural Economics Research Institute (LEI), The Netherlands

Schroeder D, Palmer C (2003) Technology assessment and the 'ethical matrix'. Poiesis Prax 1(4):295–307

Chapter 4
Ethical Challenges for Humans Using Traditional and Complementary Medicines

Abstract Critics of traditional and complementary medicine assert that people are putting their trust in the hands of incompetent and poorly regulated practitioners who are deliberately misleading patients about benefits leading to risks to health, and even lives. Injury or infection from acupuncture needles, severe injuries from osteopathy or chiropractic manipulation, or rejection of effective conventional medicine, are just some of the threats described for humans who use traditional and complementary medicine. The two most commonly cited safety concerns are examined: adverse drug reactions from herbal medicine and the potential for harm from delayed access to conventional care when using homeopathy. Through examination of real world scenarios and data, discrepancy between the rhetoric and the reality of these contentious issues is revealed.

Keywords Adverse drug reactions · Adverse events · Herbal medicine
Homeopathy · Patient safety · Pharmacovigilance

4.1 Adverse Effects in Medicine

The ethical challenges related to the human use of T&CM in the previous chapter are both numerous and diverse. Issues pertaining to wellbeing, autonomy and justice all warrant attention, but there are far too many to examine in detail. Hence, my analysis in this chapter is restricted to the most commonly cited concerns in the literature, namely issues related to safety.

Conventional medical protocol distinguishes between two different types of adverse effects from medical interventions: adverse drug reactions (ADRs) and adverse events (AEs).

The World Health Organization (WHO) define an ADR as, 'a response which is noxious and unintended, and which occurs at doses normally used in humans for the prophylaxis, diagnosis, or therapy of disease, or for the modification of physiological function' (WHO 1972). ADRs are characterised by the suspicion of a causal relationship between the drug and the occurrence, and are judged as being at least

© The Author(s), under exclusive license to Springer Nature Switzerland AG 2018   41
K. Chatfield, *Traditional and Complementary Medicines: Are they Ethical
for Humans, Animals and the Environment?* SpringerBriefs in Philosophy
https://doi.org/10.1007/978-3-030-05300-0_4

possibly related to treatment by the reviewing health professional (UMC 2013). The term ADR is often used interchangeably with 'side-effect' which is commonly used to describe any unwanted or unintended effect of a drug occurring at normal dosage. For example, penicillin can cause allergic reactions such as rashes, hives, itchy eyes, and swelling of the lips, tongue or face. In rare cases the allergy can be severe enough to cause anaphylactic shock, a dangerous condition that can be life-threatening (WebMD 2018).

In contrast, an AE involves iatrogenic harm, and can be defined as, 'any negative or harmful occurrence that takes place during treatment, that may or may not be associated with a medicine' (UMC 2013). AEs may take the form of temporary or permanent injuries caused by poor medical management or medical errors arising from actions or omissions, and are not due to underlying disease; nor are they expected outcomes of treatment (Tsang et al. 2013). AEs might result from provision of the wrong drug, an infection that is transmitted during investigation or surgery, or failure to act upon symptoms that could indicate the presence of a serious complaint.

For this analysis, I distinguish between ADRs and AEs in T&CM, and consider ADRs as harmful effects caused by a T&CM product and AEs as harm that arises as a consequence of malpractice, negligence or omission. Thus, AEs may occur as a result of misdiagnosis, lack of referral or delayed access to effective care, for example. In my consideration of each of these two classes of adverse effects, I draw, in the main, upon challenges that have been levied at one particular form of T&CM in each case. I consider the potential for ADRs with respect to herbal medicine and for AEs I examine such cases in relation to homeopathy. This approach has been chosen as these two forms of T&CM are by far those most commonly associated in the published literature with ADRs and AEs, respectively. Additionally, while I undertake this analysis from a UK perspective, with reference to UK procedures and regulations, in the main, this is reflective of what happens across Europe and is of relevance in the global environment. I begin with consideration of safety procedures in the conventional healthcare system.

4.2 Safety in the Conventional Healthcare System

Consideration of safety in the conventional healthcare system is of value as a starting point as this provides an established context for the examination of, and comparison with, safety issues arising in the use of T&CMs. The processes involved in the monitoring of ADRs and AEs in conventional healthcare in the UK overlap in many respects as both rely heavily upon reporting systems. Additionally, it is not always easy to determine whether an adverse effect is the result of an ADR or an AE, as the identification of an ADR involves an assessment of causality between the harmful effect in the patient and the medicinal product, and it is not always possible to determine this with confidence. Nonetheless, there are some distinct procedures in place for each.

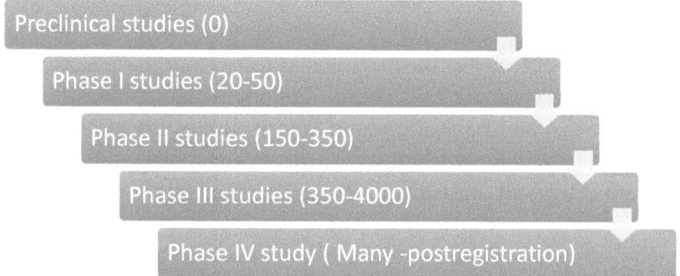

Fig. 4.1 Stages in the development of pharmaceutical products

4.2.1 Monitoring Adverse Drug Reactions

Fairly rigorous mechanisms are in place for the ongoing detection and prevention of ADRs, collectively termed *pharmacovigilance,* and the implementation of pharmacovigilance around the world is essential for controlling the incidence of ADRs. Pharmacovigilance is defined by the WHO (2004) as the science and activities relating to the detection, assessment, understanding and prevention of adverse effects or any other medicine-related problem. Today, the procedures involved in the pharmacovigilance of medications are generally stringent, but this was not always the case. Tightening of controls relating to drug development and drug use became a high priority in healthcare after the negative effects of thalidomide were revealed in the early 1960s in a public and high profile manner, leading to calls for greater regulation of pharmaceuticals (Heaton 1994). There are various stages involved in the clinical development of pharmaceuticals and information related to safety concerns can be collected and assessed at each stage. Following the preclinical stage, increasingly large numbers of humans are exposed to the proposed drug as detailed in Fig. 4.1.

Figures relate to the approximate number of human participants involved at each stage.

Preclinical Studies

No humans are involved in this initial stage, and the approach usually consists of in vivo animal laboratory studies and/or in vitro experiments.

Phase I Studies

These are the first studies of a drug on humans. They involve an initial evaluation of the effects on a small number of research participants (between twenty and fifty individuals), usually healthy volunteers, in order to assess toxicity, tolerance and aspects of pharmacokinetics.[1]

---

[1] The study of the absorption, distribution and elimination of a pharmaceutical in the body.

Phase II Studies

It is at this stage that efficacy of the drug is first tested and these studies are usually the first conducted on patients. This phase often involves the investigation of dose-response relationships and bioavailability.[2] These studies are exploratory in nature and are often used to decide which dose(s) or treatments merit further investigation. Sample sizes are often still quite low (150–350 participants).

Phase III Studies

In this phase, a formal comparison of the new drug at the dose(s) chosen from Phase II is undertaken with the standard current treatment or placebo. It involves considerably more patients than Phase II (250–4000). The primary objective is to determine whether a drug, for which likely effective and tolerable doses have been established in phases I and II, is suitable for registration, based on its efficacy and safety.

Phase IV Study

This phase involves monitoring the long-term effects of a drug, usually after registration, through its use and administration in the real world. In particular, investigation of efficacy in different populations can be undertaken and information on rare adverse effects can be gathered that have not been revealed at previous stages, in large part due to the much greater population size (WHO 2004).

One of the functions of clinical studies is to determine common adverse effects and these will normally be evident from studies undertaken during phases I–III. However, these studies are conducted on a limited number of people who have been carefully selected and run over a finite amount of time. This presents several obstacles to the detection of certain ADRs, in particular:

- Those that appear a long time after drug exposure, such as cancer, or those that develop after chronic use, such as the long-term ADRs of oral contraceptives that can take years to develop (Sultana et al. 2013).
- Less common or rare ADRs that are not revealed in the drug trials because drugs are rarely tested on more than 5000 people before release (WHO 2004).
- Those that appear in people with different characteristics from the participants used in the trials, such as people of different ages, those with multi-morbidity, or those simultaneously consuming other medications (Scott and Thompson 2014).

Consequently, many ADRs are only detected after a product has been used by a large number of people over a long period of time, and it is essential that new treatments are monitored for their effectiveness and safety under real-world conditions following their release. This is particularly important for gathering information about use by groups who are underrepresented or omitted from previous stages such as the elderly, children, pregnant women and those taking other medicines. Experience has shown that the introduction of a new medicinal product often carries unknown risks.

---

[2]The fraction of a dose absorbed by the subject and the rate at which it is absorbed.

Hence, in addition to the assessments made during drug development, post-market surveillance is also needed. This may take the form of individual case reports, cohort studies, population statistics and meta-analyses (Scott and Thompson 2014). Information about ADRs can be obtained through spontaneous reports from healthcare practitioners and patients or through the application of epidemiological studies of ADRs.

In the UK, the Medicines and Healthcare products Regulatory Agency (MHRA) monitors the safety of healthcare products through the mediation of the 'Yellow Card Scheme'. This scheme relies upon spontaneous reporting by healthcare professionals or patients, and reports can be submitted for all medicines, including vaccines, blood factors and immunoglobulins, for herbal medicines and homeopathic remedies, and for all medical devices available on the UK market. Information is collected on suspected problems or incidents involving:

- ADRs (side effects),
- Medical device adverse incidents,
- Defective medicines (those that are not of an acceptable quality), and
- Counterfeit or fake medicines or medical devices (MHRA 2015).

Where deemed necessary, the MHRA then reviews the product and can take action with a view to minimising risk and maximising benefit to patients. This may include passing on the report to an international monitoring centre.

In addition to national reporting schemes, verification of a new, potentially harmful reaction often requires the collection and review of reports from other countries, and these reports must be properly assessed and validated. The WHO have been key players in establishing the international monitoring of ADRs through their programme for international drug monitoring.

Since 1978, the administration and operation of the programme has rested with WHO collaborators, Uppsala Monitoring Centre (UMC) in Sweden, who maintain the WHO global database of individual case safety reports *VigiBase*. This is the largest database of its kind in the world, with over 16 million reports of suspected adverse effects of medicines from around 150 countries (UMC 2018).

It must be emphasised that these case safety reports detail *suspected* ADRs and not actual ADRs. Before an adverse effect is classified as an ADR, the likelihood of causality between the medication and the adverse effect must be assessed. The success of pharmacovigilance is thus, to a large degree, reliant upon the assessment of causality, but this inevitably involves a subjective element of judgment on the part of the assessor. There is currently no methodology by which an assessment of causality can be undertaken in a completely objective and quantitative manner. The WHO-UMC system for assessment of causality takes into account the clinical aspects of case histories and the quality of the documentation of the observation provided.

Crucially, the relatively low number of reports received limits the effectiveness of post-marketing surveillance. It is estimated that, in the UK, less than 10% of all serious ADRs are spontaneously reported (Scott and Thompson 2014) and the UK is one of the largest contributors to international surveillance. Most ADRs are therefore unreported and hence the nature of ADRs from conventional medicine and the figures for incidence can only be approximated (Luteijn et al. 2012).

4.2.2 Monitoring Adverse Events

The detection and monitoring of AEs is even more complex than the detection of ADRs and systematic gathering of data related to AEs is a more recent phenomenon. Contemporary developments were prompted at the end of the 1990s when the Institute of Medicine (IOM) in the United States issued a wake-up call with its report entitled, 'To Err is Human' (Kohn et al. 2000). This report asserted that, in the US, there were 44,000 to 98,000 preventable deaths annually arising from medical errors in hospitals. The resulting media interest brought the issues of medical error and patient safety to the forefront of national concern. Within a very short period of time, the IOM report was credited with truly 'changing the conversation' and stimulating a broad array of stakeholders to engage in patient safety, as well as motivating hospitals to adopt new safe practices (Leape and Berwick 2005).

Like the USA, the UK quickly adopted new procedures designed to improve safety. Currently, statutory patient safety functions fall under the remit of *National Health Service (NHS) Improvement*; the name, presumably, acknowledging that there are improvements to be made. In spite of major efforts, the health service in the UK faces significant problems. A recent report (Elliott et al. 2018) estimated that 237 million medication errors occur at some point in the medication process in England per year, of which, 66 million are potentially clinically significant. The same report estimates NHS costs of definitely avoidable ADRs of £98.5 million per year, causing 712 deaths, and contributing to a further 1708 deaths. Significantly, the authors discovered that error rates in the UK are similar to those in other comparable health settings such as the US and other countries in the EU.

According to the WHO, as many as one in ten patients is harmed while receiving hospital care, arising from errors or adverse events. For example, of every one hundred patients hospitalised at any given time, seven will acquire health care-associated infections even though many of these can be prevented by simple control measures such as appropriate hand hygiene (WHO 2014). Globally, there are approximately 43 million patient safety incidences every year and medication errors cost an estimated 42 billion USD annually (WHO 2018).

It is clear that safety assurance is a significant challenge for those in conventional healthcare in spite of the existence of clear infrastructures and procedures for the identification and prevention of ADRs and AEs. Most practices of T&CMs fall outside the procedures of the established healthcare system. They use products that have generally not been tested in the same manner as those that pharmaceutical products are subjected to, and there is a lack of coherent procedures for reporting

ADRs and AEs. This shortcoming in clinical governance has led to accusations of considerable risk of harm to patients (Hunt and Ernst 2010). In the following sections I examine whether these accusations are justified and, if so, how they might be addressed.

4.3 Safety in T&CM

Concerns about ADRs from T&CM products are primarily focussed upon herbal products, whether prescribed by a practitioner or self-medicated. Herbal products have been linked with risks of drug/herb interactions, allergic reactions, toxicity, drug contamination, interference with laboratory tests and overdose (Posadzki et al. 2013a).

Aside from the potential ADRs from herbal medicine products, there are also potential risks from AEs, of a more intangible nature. Over the past twenty years, increasing numbers of concerns about actual or potential AEs for people using T&CM treatments have been published. Those revealed in the literature include misdiagnosis, delayed diagnosis, and failure to use effective treatments with resultant unnecessary morbidity or mortality. In addition, there is a potential for practitioners to cause harm through inappropriate advice, the provision of misleading information, a lack of professional boundaries or poor standards of care.

Nowadays, AEs are broadly classified as either *errors of commission* (unintentionally doing the wrong thing), or *errors of omission* (unintentionally not doing the right thing) (Garrouste-Orgeas et al. 2012). For example, incompetence on the part of the practitioner could lead to serious AEs from acts of commission like trauma and injury from acupuncture, chiropractic or osteopathy (Rajendran et al. 2012); herbal medicines may be misprescribed (Vickers et al. 2001). However, as more information becomes available, it is becoming clear that serious AEs associated with acts of commission are not common in the UK. For instance, a systematic review of thirty-nine clinical studies of manual therapy (including chiropractic and osteopathy) revealed that *minor* AEs are common after treatment, with over half of patients experiencing short-lived reactions, but that the risk of *major* AEs was found to be very low; lower than it would be from conventional medications (Carnes et al. 2010). Similarly, two prospective safety surveys from the UK, which were based on more than 66,000 acupuncture sessions, did not report any serious AEs (MacPherson et al. 2001; White et al. 2001).

By far the most common accusations of risk from T&CM concern acts of omission rather than commission. Here, concerns arise from the potential for rejection of, or delay in access to, conventional care leading to serious consequences that might otherwise have been avoided. An underlying assumption here being that conventional care is effective while T&CM is not. In addition, it is assumed that practitioners of T&CM may fail to recognise serious conditions and/or interfere with a conventional treatment approach through contradictory advice about conventional medications and immunisation (Ernst 2009).

Children can be put at risk when parents opt for T&CM treatments (Lim et al. 2011) as was tragically highlighted by the deaths of four children in Australia in 2010, where the cases were reported to relate to a failure to use conventional treatment. One case involved an infant of eight months who was admitted with malnutrition and septic shock following naturopathic treatment with a rice milk diet for 'congestion'. Another death was of a child who was prescribed anticoagulants following pulmonary emboli, but who was treated with an unspecified complementary medicine instead. The child died following complications relating to a pulmonary infarction.

Further concerns about T&CM for human users arise from the wide variation in standards of education, training and regulation, leading to questions about the clinical competence of practitioners (Nissen et al. 2013). Concerns about the safety of practitioners apply to all forms of T&CM and include issues such as: working within bounds of competence, the provision of reliable information, and integration with other forms of health care.

While the reported number of cases of ADRs and AEs from T&CM is tiny in comparison to those associated with the use of conventional medicine, there is undoubtedly a potential for harm from use of T&CM, just as there is from conventional medicine. Hence, the subject is worthy of scrutiny and in the following section, attention is turned to ADRs through examination of the case of herbal medicine.

4.4 Adverse Drug Reactions and Herbal Medicine

MHRA 28th February 2018
Press release

Natural doesn't mean safe—herbal medicines found to contain steroids

The Medicines and Healthcare products Regulatory Agency is warning people who may have purchased a "natural" Chinese herbal medicine, Yiganerjing Cream, as a treatment for skin conditions to stop using it immediately as it has been found to contain an undisclosed steroid and two antifungal ingredients.

MHRA officials have been acting to stop the sale of this cream and have had it withdrawn from many websites and online market places but people may have purchased it in the past and still be using it.

In the UK, reports of suspected ADRs associated with herbal products are reviewed by the MHRA who will issue warnings like the above via the government website when there is a perceived risk to health. It is clear from examples like this that there is a potential for harm from unlicensed and unregulated products containing herbal medicine especially from poor quality, adulterated or contaminated products. Additionally, even with high quality products, there are risks of overdose, side effects, drug/herb interactions and interference with biological tests. Many plants are potent or toxic and there is typically far less safety data available for herbal products than would be required for conventional medications (Werner and Soghomonyan 2014).

Herbal treatments can be prescribed on an individual basis following consultation with a trained medical herbalist, but many herbal products are also available over-the-counter in pharmacies, health food stores and supermarkets, either as single herbs or in compound formulations. The market for T&CM over-the-counter products is large. According to Future Market Insights (2017), the global market for herbal medicinal products is increasing and expected to be valued at more than 130 Billion US dollars in 2017.

A significant number of people hold the mistaken assumption that herbal products are safe because they are natural (Okoronkwo et al. 2014). However, herbal medicines have pharmacological effects, just like synthetic pharmaceuticals, including the potential for ADRs. Additionally, a vast array of products containing herbal medications that are sold online are completely untested and unregulated (Bhagavathula et al. 2016). Unlike pharmaceutical drugs, most herbal medicines are not taken through the same phases of development from preclinical to phase III studies before release onto the open market. In fact, most herbal medicines have been in use for substantial periods of time, perhaps even hundreds or thousands of years, before they are subjected to clinical studies, if at all. Figure 4.2 provides an overview of the stages of development. *These stages are representative of current day development and not the original indigenous developmental processes.*

There are estimated to be more than 70,000 species of plants used in traditional medicines around the world (Chen et al. 2014), but only a tiny fraction have been subjected to rigorous testing under controlled conditions for efficacy and safety. It is only in the last thirty years or so that the risks involved in the use of herbal medications has demanded more than cursory attention and the clinical studies of herbal medicines that have been undertaken vary greatly in quality and value (Posadzki et al. 2013a).

The inclusion of herbal medicine in the monitoring processes for ADRs, both nationally and internationally by the WHO, has helped to reveal some of the potential risks involved with using herbal medications. In the UK, reports to the MHRA are considered each year by the Herbal Medicines Advisory Committee (HMAC) and in 2016, this amounted to a total of 89 reports, of which, 55 were considered serious

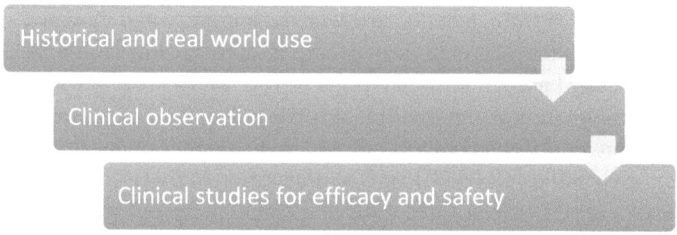

Fig. 4.2 Stages in the development of herbal medicines

(HMAC 2017). The figures for the reporting of suspected ADRs in the UK are very low, but when taken together with reports from other countries they become more significant.

In the period 1968–1997 a total of 8985 individual ADRs involving the use of herbal medicines were reported (Farah et al. 2000). At that time anaphylaxis, including anaphylactic shock, was the most frequently reported critical adverse reaction at 67 out of 2487 reports (2.7%). Pruritus (itching skin) was the most reported non-critical reaction with 324 out of 2487 reports (13%), followed by diarrhea, which was often associated with the use of single ingredient herbal medications, and comprised 109 cases (4.4%) (ibid.).

By December 2010, there were 12,679 suspected case reports in the WHO database where purely herbal substances were involved and 21,951 reports that included both herbal and non-herbal substances (Farah 2011). The most commonly reported critical ADRs from use of herbal products being drug abuse (630), drug dependence (274) and hepatitis (263). The most commonly reported herbal medicines associated with suspected ADRs were, *Cannabis sativa* (1057), *Ginkgo Biloba* (960) and *Hypericum perforatum* (713).

These figures provide an overall impression of the types of *suspected* ADRs, as well as the medicines most commonly associated with them, but they do not provide details about subsequent assessments of causality or potential mechanisms of harmful effect. In addition, they are not helpful for distinguishing between types of ADRs such as toxic effects of the herbal medicine, the potential for overdose or allergic reactions, contamination of the product, or adverse interactions with conventional drugs. Instead, they represent the early stages of an effort for global pharmacovigilance of herbal products that still has a long way to go in addressing the primary ongoing challenges. Four of these challenges are outlined below, namely hepatoxicity, herb/drug interactions, contamination of herbal products and the misidentification of herbs.

4.4.1 Herbal Medicine and Hepatoxicity

It may seem reasonable to assume that the extensive use of herbal medicines over long periods of historical duration should have inevitably revealed implications for their safety. Information about who might be at risk, how much to take, and potential side effects can be gleaned from years of real world experience and guidance developed accordingly. This type of information evolves as it is passed down through generations and is vital to locally based, traditional forms of T&CM. However, when products are removed from their cultural and traditional roots and applied in completely different environments, under different conditions, and in different formats, the consequences of use are unpredictable (WHO 2013). One illustration of this is the case of kava kava (*Piper methysticum*). For over three thousand years, the people of the South Pacific Islands have been using kava kava for medicinal, religious, and social purposes. There is a great local respect for this plant which holds deep cultural significance for these peoples (Cawte 1985). In Fiji, for example, the kava kava ceremony often accompanies important occasions when the root of the plant is chewed or pounded, and then mixed with water, strained and drunk. The effects of the kava kava are sedative and have been likened in some respects to alcohol (Thompson et al. 2004). Following reports of medicinal benefits, the leaves and the root of the kava kava plant were used to make herbal products that became popular in many other parts of the world for the treatment of anxiety, tension, insomnia and restlessness. Clinical studies backed up the efficacy claims for the use of kava kava in the treatment of anxiety (Connor et al. 2006).

In 2001, concerns were raised about the safety of kava kava after a number of cases of acute liver failure were reported, which led to restrictions, regulations and outright bans in some countries. While kava kava has the potential for beneficial properties, it appeared that it also has the potential for hepatotoxic effects in certain individuals (Wooltorton 2002). Subsequent analysis has indicated that the majority of these case reports are probably not connected to the intake of kava kava (Teschke et al. 2013), but still the potential for hepatotoxic effects has not been ruled out and internal use of the product remains banned in the UK. More research is needed to identify the precise requirements for the safe use of kava kava, despite its extensive use in the South Pacific.

Kava kava is not the only herbal medicine to be associated with hepatoxic effects. In their review of reports of serious hepatotoxic events, Abdualmjid and Sergi (2013) examined 254 reports of hepatoxicity involving herbs alone or in combination with other drugs. Twenty-seven herbs were subsequently identified as having a recognised potential for hepatoxic effects and the authors warned that, as the use of natural medicine increases, so will the risk of liver toxicity. The potential risk of damage to the liver from herbal medicines is now widely acknowledged. These include single herbs such as kava kava as well as compounds with often ill-defined ingredients (Stickel and Shouval 2015).

4.4.2 Herb/Drug Interactions

Some herbal medications can lead to harm in people who are simultaneously taking herbal medicines and conventional drugs, due to adverse interactions between the herbal medicine(s), the drug(s) and the human body. They can cause problems if they interfere with the movement of a drug through the body, or altering the effect of the dosage on the body in an ampliative or inhibitory manner (Le 2014). However, there is considerable debate about the extent to which herbal medications can interfere with conventional drugs, which herbal medicines are risky, which drugs most affected, the problems caused, and the mechanisms of action. Many of the studies investigating potential interactions are conducted using in vivo animal studies or in vitro laboratory experiments, and the results subsequently extrapolated for predictions of effects in humans. Within the herbal medicine community there are feelings of unjust assessment, as Josephine Briggs, director of the NCCIH points out:

> There are 11 major drug interactions with coffee, yet doctors don't tell patients not to drink
> coffee based on possible interactions! A lot of the fears about herbs are not founded on good
> meaningful accurate data (Briggs quoted in Tweed 2015).

However, as the number of studies increases, the picture is becoming clearer and there is some cause for caution with at least a limited number of herbal medications. Following their analysis of forty-six systematic reviews of potential drug-herb interactions, Posadzki et al. (2013c) concluded that the majority of herbal medicine products are not associated with severe herb/drug interactions. Serious interactions were noted only for *Hypericum perforatum* and *Viscum album* and moderately severe interactions were noted for *Ginkgo biloba, Panax ginseng, Piper methysticum, Serenoa repens* and *Camellia sinensis*. The most commonly interacting classes of drugs were found to be antiplatelet agents and anticoagulants. The most serious interactions resulted in a range of conditions that included transplant rejection, delayed emergence from anaesthesia, cardiovascular collapse, renal and liver toxicity, cardiotoxicity, and death. However, the authors warned that the poor quality and the scarcity of the primary data prevent firm conclusions and called for further investigation. Gallo et al. (2014) also found problems with antiplatelet agents and anticoagulants in their study of 478 preoperative patients in Italy, as well as antihypertensive and central nervous system agents. Of the 478 participants in the study, forty-two were identified as being affected by a herb/drug interaction.

4.4.3 Contamination of Herbal Products

A further potential for harm arises from the use of herbal medicines which may be contaminated, either deliberately or unintentionally, as is clear from the aforementioned MHRA warning on the UK government website. In a review of previous systematic reviews investigating the adulteration and contamination of herbal medicinal products, the most common contaminants were named as: dust, pollens, insects, rodents, parasites, microbes, fungi, mold, toxins, pesticides, toxic heavy metals and/or prescription drugs. A very wide range of ADRs from contaminated products were reported, with the most severe resulting in cases of: meningitis; multi-organ failure; arsenic, lead or mercury poisoning; malignancies or carcinomas; renal or liver failure; cerebral oedema; and coma or death (Posadzki et al. 2013b). The systematic reviews included reports about herbal products from all over the world and concluded that contamination was most common in traditional Indian and Chinese remedies.

There are two types of herbal products most commonly associated with deliberate adulteration: weight loss products and sexual performance enhancers. For example, Gilard et al. (2015). analysed 150 supplements marketed as natural products for sexual performance enhancement claiming to contain only natural compounds, plant extracts and/or vitamins. Of the 150 products analysed, 61% were adulterated with phosphodiesterase-5 inhibitors (the active ingredient in Viagra), with 25% of these at a dosage higher than the maximum recommended dose. Of the whole batch, only 31%

of the samples were considered true herbal/natural products. Furthermore, a follow-up of several products over time revealed that some manufacturers make changes to the chemical composition of the formulations so it is virtually impossible for a member of the public to have confidence in the manufacturer's description (ibid.).

4.4.4 Identification of Herbs and Herbal Ingredients

As aforementioned, the number of plants used for medicinal purposes around the world is estimated at more than 70,000 species (Chen et al. 2014). Given this vast number, and the wide variation in species that are sourced, as well as variable standards of quality control and regulation of products, it is not surprising that a number of safety-related issues have emerged because of the inaccurate identification of herbal materials.

One of the most widely publicised incidents of ADR from mistaken identity of a herb was reported in Belgium in the 1990s, when at least one hundred women were believed to have developed progressive renal failure after adhering to a weight-loss regimen that included the use of Chinese herbs. Suspicion that the disease was due to the recent introduction of Chinese herbs in the slimming regimen was reinforced by identification of aristolochic acids in the slimming pills. These compounds are known to cause nephrotoxicity and carcinoma of the upper urinary tract (Grollman et al. 2007). In depth analysis of the pills revealed that the prescribed Chinese herb, *Stephania tetrandra* was not present in the pills at all, and in fact this had inadvertently been replaced by another Chinese herb, *Aristolochia fangchi*, in the herbal extracts used in Belgium and in France (Vanherweghem 1998). Similar cases have been observed throughout the world and, in the majority of known cases, the cause of the problem has been the inaccurate identification of the plant ingredients (Chen et al. 2014). The correct identification of medicinal plant ingredients is therefore essential for their safe use.

4.4.5 Summary of the Challenges

Herbal medications have pharmacological effects on the body, just as conventional medications, but they are generally poorly researched in comparison with conventional medicines in terms of their efficacy and safety. Conventional drugs must undergo extensive testing in clinical studies, but therapeutic knowledge about herbal medicines is gleaned in a different manner. Most of what we know about herbal medications is derived from experience of their traditional and local usage, in some cases over hundreds or thousands of years. According to the WHO (2013), an increase in reports of ADRs associated with herbal products reflects a growing awareness that natural products may also cause harm.

However, the public are often unaware of the potential for harmful effects from herbal products, assuming that the term 'natural' implies safety. Self-medication without appropriate knowledge can lead to harmful herb/drug interactions, overdose or toxicity. No international standardisation guidelines for processing, manufacturing and marketing of herbal products exist (Kim et al. 2014) and herbal products may be contaminated with conventional medications, toxins, or ingredients that have been misidentified.

4.5 Addressing the Ethical Challenges

In light of these challenges there are three broad requirements for reducing ADRs from herbal products. First, more safety data is needed to assess toxicity levels, appropriate dosage, the likelihood of side effects and the potential for adverse interactions with other medications; secondly, herbal products require quality control and licensing on a global scale; and thirdly, the general public and health professionals need access to the most up to date and reliable information about potential ADRs in the context of herbal medicines.

Worldwide scientific study of herbal medicines is now increasing, together with improvements in the sharing of information between regulators, and this is being driven, to a large degree, by increased public demand for herbal products in high income countries and guided by initiatives from the WHO. Safety data is growing thanks to improvements in reporting systems for ADRs on national and international levels.

For accurate analysis of reports of ADRs from herbal products, it is vital to know the exact scientific name of the plant, the part of the plant that was used and the name of the manufacturer. The UMC have been collaborating with the Royal Botanical Gardens at Kew for a number of years to improve standards in identification of herbal materials. Currently, Kew's Medicinal Plant Names Service (MPNS) is seeking to capture all names (common and scientific) for herbal medications from a broad range of sources. Each name is mapped to Kew's botanical references which captures all synonyms for each plant. MPNS is thus able to reliably count species and makes links between different names for the same plant. By May 2017, MPNS contained 384,000 unique names for 26,000 plants (UMC 2018).

In the UK, the MHRA launched a registration scheme for herbal medicines in 2005 with the aim of improving the safety of herbal products that are available over-the-counter. Known as the Traditional Herbal Registration Scheme (THR), this scheme dictates which herbal medicines are available over-the-counter and sets specific standards for safety and quality. Products approved by the scheme have a 'THR number' on their labels and registration is only available to those medicines used for minor health conditions where medical supervision is not required (for example, colds and coughs). To be eligible for registration, the indication (the medical condition the product has traditionally been used to treat) must be agreed under legislation. For agreement of this legislation, evidence that the herbal medicinal product has been used traditionally to treat the stated condition for a minimum of thirty years (fifteen years of which must have been in the European Union) is required. Herbal remedies which are administered by a qualified herbal practitioner, following a consultation, do not require a medicines licence, but may be subject to certain restrictions. For example, some herbs, such as aristolochia and kava kava, are deemed toxic and are banned for use in the UK under any circumstances (MHRA 2014).

On a global level, the WHO is taking a proactive role in driving forward the development of policies and regulation of herbal products. According to the WHO, most Member States now regulate herbal products, although new regulations are continually being developed, updated and implemented (WHO 2013). As the market is now truly international, with products often being made in a country other than that in which they are sold, it is acknowledged that it may be a challenge to ensure products are safe and of high quality. In order to address this issue, Member States and regulatory agencies are cooperating and learning from each other's experiences. Some regional bodies have been working on harmonising regulations on herbal medicines and other herbal products within their region (ibid.). It is therefore anticipated that over time levels of safety of T&CM products will improve.

4.5.1 Ethical Analysis: Adverse Drug Reactions and Herbal Medicine

For ethical analysis of issues pertaining to the human use of herbal medicines, and the associated impacts upon wellbeing, the potential for harm in the form of ADRs needs to be carefully considered alongside the potential for benefit. Examination of the ethical challenges has revealed a potential for harm from the use of herbal medications but, in the UK at least, steps have been taken to minimise this potential. Over-the-counter products must be licensed and there are restrictions upon the practitioner prescription of some herbs that have been deemed risky. In addition, reporting procedures are in place for collection and analysis of safety data that can help to improve safety in the future.

The safety data available for herbal products is generally of a different quality and nature to that for conventional medications. In some respects, the long history of use of most herbal medications provides us with more realistic safety data than is available for conventional medications. Experience of use within large and diverse populations, over extended periods of time, provides valuable real-world information, whereas conventional pharmaceuticals are typically released onto the open market having been tested on a limited number and range of individuals, over a finite amount of time. However, historical usage does not necessarily guarantee safety.

Even so, data suggests that the number of serious ADRs from herbal medications is extremely small, especially when compared to the number from conventional medications. In their review of fifty systematic reviews of adverse effects of single herbal medicines Posadzki et al. (2013a) found that severe adverse effects were noted for only four herbal medicines: *Herbae pulvis standardisatus* (belladonna*), Larrea tridentate* (chaparral*), Piper methysticum* (kava kava), and *Casia senna* (senna), with three of these having been identified as potentially hepatoxic: chaparral, kava kava and senna (Abdualmjid and Sergi 2013). However, moderately severe adverse effects were noted for a further fifteen herbal medicines and mild effects for a further thirty-one herbal medicines.

Ernst, reported that even though herbal medicines are not devoid of risk, they could still be safer than synthetic drugs (Ernst 2003). In support of this view, Ernst quotes Linde et al. (1996) who found that the herbal antidepressant St John's wort had only half the rate of adverse effects when compared with conventional antidepressants and Schulze et al. (2003) who claimed that even though kava kava had been banned in the UK, it was probably still safer than benzodiazepines.

The potential benefits of herbal medicines are not limited to their clinical effectiveness; other benefits can stem from their high acceptance by patients and their relatively low costs. Patients worldwide seem to have adopted herbal medicines in a major way (WHO 2013). In the UK, access to conventional medical healthcare is free to all citizens through the NHS yet many people elect to pay themselves for herbal products either instead of, or as well as, conventional products in an ongoing fashion. It is difficult to imagine why people would continue to pay out of pocket for treatments if they did not perceive some type of benefit. Indeed, Pirotta et al. (2014) found that people experiment with different treatments and continue to use what they perceive is most effective for them.

Evidence of efficacy may be limited for many herbal medications but acceptability for them appears to be high and associated risks are low, with one notable exception. It is possible to access herbal products via the internet that are neither licensed, nor prescribed by a practitioner, and these products are typically prone to batch-to-batch variability in composition, concentration, contamination, and purposeful adulteration (Stickel and Shouval 2015). If people choose to use unlicensed medications obtained via the internet (especially weight loss products and sexual performance enhancers) then it would appear they are putting themselves at much greater risk of suffering an ADR.

4.5.2 Recommendations for the Avoidance of Adverse Drug Reactions from Herbal Medicine

It is not possible to assess accurately all of the potential benefits of herbal medications from the information that is currently available, but it is possible to identify factors that may reduce the potential for harm.

For people using herbal products, the risk of experiencing ADRs from herbal preparations can be minimised if they:

- Inform their healthcare practitioners (conventional and T&CM) when taking herbal medications alongside conventional medications.

- Adhere to the instructions for dosage either on the packet for over-the-counter products or as provided by the practitioner for prescribed products.
- Use a reporting scheme for any suspected adverse reactions.
- Never buy unlicensed herbal medicinal products via the internet.

For practitioners, the risk of harm to their patients can be minimised if they:

- Keep up to date with safety data concerning the potential for ADRs from herbal medicines.
- Inform patients of the potential for ADRs (where known).
- Inform patients about what to do if they suspect that they are experiencing an ADR.
- Make full use of reporting schemes for suspected ADRs.
- Ensure that patients understand how to use their prescribed herbal medications.

For patients who are concurrently using conventional medication, they should:

- Take the relevant steps needed to avoid herb/drug interaction, including referral back to the conventional healthcare professional when appropriate.

Thus, there are clear steps that can be taken to reduce the potential for harm from ADRs associated with herbal medicine. The potential for harm from AEs associated with T&CM, however, requires a completely different strategy. The risks of AEs associated with the use of T&CM are considered next through the example of homeopathy.

4.6 Adverse Events and Homeopathy

By far the most common accusations of risk from T&CM concern acts of omission. The concern that practitioners of T&CM may fail to recognise serious conditions and/or interfere with conventional treatment is commonplace (Gilmour et al. 2011a) and this leads to the accusation that, if people opt to use T&CM treatments, the rejection of, or delay in access to, conventional care can have serious consequences (Curtis and Gaylord 2005). While this may be a concern for all types of T&CM, in the current literature it is most frequently levied at homeopathy. However, there are no more recorded cases of AEs from homeopathy than for other types of T&CM; the primary reason why homeopathy is the most frequent subject of this challenge appears to stem from an underlying challenge that homeopathy poses for conventional science.[3] Homeopathy is often equated with being purely a placebo treatment

---

[3]Homeopathic remedies are prepared from a very broad range of animal, mineral and plant substances that are serially diluted in water and ethanol. Between each dilution, the solution is succussed (shaken vigorously). Homeopaths believe that this method of preparation increases the potency of the remedies and lessens the potential for side effects. However, the process of dilution commonly exceeds the point at which any molecules of the original substance might be expected to remain in solution (as relating to Avogadro's constant), leading to the accusation that the remedies can be no more effective than water.

because there is allegedly *no plausible explanation* for the mechanism of action of homeopathic remedies (Sehon and Stanley 2010).[4] The idea that people may opt (unwittingly) for a treatment that is no better than placebo in place of 'effective' conventional medicine raises many ethical concerns because, 'patients who do not seek medical advice from properly qualified doctors run the risk of missing serious underlying conditions while they have their symptoms treated with a placebo' (Science and Technology Committee 2010).

Homeopathy is a controversial treatment and arguments can be vociferous between its opponents and its supporters. In 2005, when the Lancet published a damming meta-analysis of clinical trials of homeopathy compared with clinical trials of conventional medicine (Shang et al. 2005), it was accompanied by a short, anonymous editorial entitled 'The end of homoeopathy' that called for 'doctors to be bold and honest with their patients about homeopathy's lack of benefit' (Editorial 2005). In response, the late Dr. Peter Fisher, Clinical Director at the Royal London Hospital for Integrated Medicine and Physician to Her Majesty the Queen was highly critical of both the meta-analysis and the accompanying Lancet editorial commenting that, 'Regrettably, this attack will only widen the divisions. The way forward is open, transparent science, not opaque, biased analysis and rhetoric' (Fisher 2006).

Assessment of the research evidence for the efficacy of homeopathy paints a confusing picture for the general public as contradictory conclusions are frequently drawn. Clearly, the issue of whether or not homeopathy is more than placebo is not one that can be resolved within the scope of this book. What is of concern here is whether or not the use of homeopathy exposes people to a risk of AEs and, if so, how these can be minimised. This will be the focus of the subsequent analysis.

4.6.1 Known Cases of Adverse Events from Use of Homeopathy

There have been reports of AEs with fatal consequences for users of homeopathy. Australian barrister Ian Freckelton SC, in his alarmingly entitled article, *Death by Homeopathy*, reports on four legal cases from different countries, where outcomes were fatal (Freckelton 2012). Of the four cases, one is from England, one from India and two from Australia.

The English case involved Dr. Marisa Viegas, a general practitioner with an interest in homeopathy who purportedly advised one of her patients against taking conventional medications and instead suggested some homeopathic remedies (Chancellor 2007).

---

[4]This view has been challenged by a small number of scientists who have proposed explanations that are consistent with current scientific thinking; for example, Rutten, L., Mathie, R.T., Fisher, P., Goossens, M. and van Wassenhoven, M., 2013. Plausibility and evidence: the case of homeopathy. *Medicine, Health Care and Philosophy, 16*(3), pp. 525–532.

In India a claim for negligence was brought against the homoeopath, Dr. Patel, in 1996 who had treated a person with strong antibiotics on the basis of his own diagnosis without conventional testing. Dr. Patel was a qualified practitioner of homeopathy but he was not qualified to prescribe conventional medications (Supreme Court of India 1996).

One of the two Australian cases involved an infant who died at the age of nine months from septicaemia following ineffective treatment for chronic eczema and associated malnutrition. The girl's father, Thomas Sam (a homoeopathic practitioner) and her mother were prosecuted for manslaughter because they had repeatedly ignored advice to seek conventional medical help for their daughter (Smith et al. 2013).

The second Australian case concerns a woman who died in 2005 as a result of metastatic bowel cancer. It was reported that her homeopath, Mrs. Scrayen, had attempted to dissuade her from having surgery during her homeopathic treatment, threatening not to treat her if she elected to receive surgery (Hope 2010).

Each of these cases could be analysed from a variety of legal, ethical, or cultural perspectives as well as from a medical viewpoint and they have indeed received due attention. For the purpose of my analysis I simply wish to highlight that there are cases that demonstrate AEs associated with the use of homeopathy. One of these cases (Patel) relates to an act of commission and the other three relate to acts of omission in which delayed access to conventional care resulted in fatal consequences. In one of these three cases (Viegas), the practitioner was a qualified conventional general practitioner, and therefore, it can be assumed there was no delay in access to conventional diagnosis. Hence, there are two actual cases[5] (Scrayen and Sam), which support the challenge that homeopathic treatment (by a non-medically qualified practitioner) can result in serious harm from the delay in access to conventional diagnosis and treatment.

4.6.2 Homeopathy and Mental Healthcare

Rather than replicating published examination of these well-known cases, for this analysis of the potential for AEs associated with the use of homeopathy, I will take an alternative approach. Some authors have elected to base ethical analysis of T&CM upon hypothetical case examples (Ernst and Smith 2018) but analysis that is based upon real world opinions, events and practice is more reliable. Hence, I will base my analysis upon the findings from a survey of UK homeopaths about their experiences and opinions of treating people with mental health concerns which revealed many practical and ethical challenges (Chatfield and Duxbury 2010).

In 2006, one tenth (200) of members of the Society of Homeopaths, the largest professional organisation for homeopaths in the UK, were randomly selected to participate in the survey. As well as quantitative data, open questions asked specifically

---

[5]Obviously there may be many more cases that have gone unreported.

for homeopaths' experiences and opinions on a number of issues such as what helped and hindered treatment, and asked them to identify any challenges that they may have encountered. Responses to the questionnaire were anonymous.

Findings suggested that homeopaths treat patients with mental health conditions on a regular basis. The most commonly treated conditions fell into the categories of anxiety and mood disorders. However, the results also indicated that homeopaths were treating people with a full spectrum of mental health disorders. For example, a substantial proportion (57%) stated that they had previously treated patients with schizophrenia and 78% stated that they would do so if requested. The sample of homeopaths spanned many years of experience yet no clear association was found between the number of years spent in practice and either willingness to treat each condition or the level of confidence in so doing.

Despite some differences in views expressed by the homeopaths, a number of generic concerns were raised in responses to the open questions. Concerns about safety for both patient and homeopath were frequently expressed and many specific examples of challenging situations were described. Fifteen homeopaths described situations where the patient posed an actual or potential danger to themselves or others, like this:

I broke confidentiality with a young borderline schizophrenic and explained to him that I would do this. In my judgement he was a potential danger to himself and was exhibiting behaviour close to stalking a female neighbour.

Nine of these fifteen homeopaths mentioned patients with suicidal thoughts and tendencies, as in this case:

I think the most difficult situation I have found myself in was when one of my patients telephoned me to say she couldn't carry on and was going to commit suicide. Luckily I persuaded her to contact other services who were very supportive and she was able to move onwards without ending her life.

From the short comments returned on the questionnaire, it is not possible to analyse these challenging situations; we do not know the precise nature of the events that led up to them or what followed. It is clear, however, that the conduct of the homeopath could be key in helping to avoid and/or minimise the potential for harm.

A broad range of conventional treatment strategies and conventional drugs are used in the treatment of people with mental health problems (Weich et al. 2011) and these can have a variety of effects upon a patient, both desirable and undesirable. Homeopathic practitioners in this survey raised concerns about treating people who are concurrently taking conventional medications, primarily because the drugs can make it difficult to determine the appropriate homeopathic prescription:

They are usually on very heavy chemical drugs which hinder establishing symptoms of the disease rather than of the drugs.

But there were also safety concerns related to the management of people taking conventional medications:

One bipolar patient on daily homeopathic prescription taking himself off lithium overnight without informing anyone.

This situation was of great concern to the practitioner because sudden medication withdrawal can lead to seriously damaging consequences. Additionally, the depth of mental health pathology may not be evident at first and a patient with a history of psychosis may appear to be functioning very well when first seen. However, should they decide to stop their conventional medicine, they may quickly relapse into an acute psychotic episode with serious consequences, and may require hospitalisation (Moncrieff et al. 2013).

4.6.3 Summary of the Challenges

The treatment of people with mental health concerns can result in challenging practical and ethical situations. There are increased safety concerns: patients may be particularly vulnerable, at greater risk of harm to themselves (and others), and even suicidal.

Mental health problems are common and many patients in the UK seek treatment from homeopaths for a wide range of mental health conditions, in particular, depression and anxiety (Spence et al. 2005). Homeopaths in the UK, most of whom are not medically qualified, treat people with a very broad range of conditions including a wide spectrum of mental health conditions. Homeopathic philosophy teaches that the person is an integrated whole of mind and body and as such no part should be treated in isolation (Vithoulkas 1980). Consequently, most homeopaths will view themselves as providers of healthcare for persons, rather than for particular types of ailments, and speciality training is rare.

In the conventional healthcare system, mental healthcare practitioners work predominantly as part of a team. They have particular expertise in the treatment of patients with mental health problems and training in how to deal with challenging situations. In contrast, few homeopaths have specialist training in mental healthcare and most work outside of the healthcare system, often on their own, and often from their own homes, without access to the standard support networks that are available to mental health care professionals (Chatfield and Duxbury 2010).

Many people with mental health problems are taking conventional medications and changes to their medications can have serious consequences (Moncrieff et al. 2013). Homeopaths may find themselves in situations that fall outside their bounds of competence, thus putting their patients and others, including themselves, at risk of harm. It is a major challenge for homeopaths to determine how they can assess their bounds of competence effectively (Stone 2000), while continually working with different kinds of patients, different kinds of illness, and in different circumstances.

4.7 Addressing the Challenges

In the UK, the profession of homeopathy is not subject to statutory regulation and there is no legal requirement for practitioners to be registered. However, there are several professional bodies that exist for the registration and regulation of practitioners, including the Society of Homeopaths (SoH). Members of SoH must meet certain criteria: they must hold certain qualifications, practise to a certain standard and abide by their Code of Ethics and Practice (SoH 2018). On their website are clear instructions about what to do and who to contact if people are concerned about a member's behaviour. In serious cases, members may be suspended or removed from the register. This system of voluntary regulation for homeopaths is intended to protect the public from harm, helping to ensure safe and ethical practise. Undoubtedly, regulation of this kind contributes to improved standards of professional practice (Gale and McHale 2015) but standards differ between professional organisations and practitioners are not even obliged to join a professional body.

The system of voluntary regulation that exists in the UK is not adopted in all European countries. In Europe, homeopathy, as a distinct system of healthcare, is recognised by law in Belgium (1999), Bulgaria (2005), Germany (1998), Hungary (1997), Latvia (1997), Portugal (2003), Romania (1981), Slovenia (2007) and the UK (1950). However, the laws in Bulgaria, Hungary, Latvia, Romania and Slovenia allow only medical doctors to practise homeopathy (European Committee for Homeopathy 2018). Clearly, the restriction of the right to practise to medical doctors in some countries avoids the challenge that practitioners are not qualified to diagnose serious conditions or manage conventional medications.

4.7.1 Ethical Analysis: Adverse Events and Homeopathy

For the analysis of ethical issues pertaining to the use of homeopathy for the treatment of mental health concerns, the potential for harm in the form of AEs needs to be carefully considered alongside the potential for benefit and the autonomous rights of the individual. Many individuals choose homeopathic treatment for their mental health problems even though this normally incurs an out-of-pocket expense. Additionally, many say that they benefit from it. For instance, 76% patients with depression who were treated with homeopathy at the Bristol Homeopathic Hospital, recorded an improvement in their condition, with 30% of these being markedly improved (Spence et al. 2005). These figures do not prove that it was the homeopathy that caused this improvement,[6] but nevertheless the majority of the patients believed that it helped them.

On the other hand, examination of the ethical challenges pertaining to AEs has revealed a potential for harm from the choice of homeopathic treatment. Of primary ethical concern is that problems can arise when practitioners work outside their bounds of competence. Practitioners who work alone, without training or experience in working with people with complex problems can be faced with challenging issues that extend beyond their normal scope of practice.

---

[6]This was an observational study and the lack of a control group means that cause and effect cannot be established. Homeopathic treatment typically involves a long consultation as well as the prescription of homeopathic medications. The improvements could have been due to either of these factors, both of these factors or something else.

Of vital importance to patient safety is that practitioners are equipped with the appropriate knowledge and skills. In order to minimise the risk of harm, practitioners must be able to recognise situations that are beyond their competence and act accordingly. Hence, careful consideration needs to be given to the issue of competence before undertaking the treatment of patients with severe or complex illness. This should be viewed as good practice, for in such circumstances it would be unethical to continue to treat the patient regardless. However, this is not a straightforward process. People with a very wide range of conditions seek help from homeopaths and the depth of their pathology is not always obvious when they first arrive.

An important consideration for the homeopath is the conventional medical treatment that the patient is receiving. As shown in the aforementioned survey, patients can, mistakenly, develop the idea that they can cease all other medication when they begin homeopathic treatment (Chatfield and Duxbury 2010). The analysis of this issue is clear: homeopaths can be considered experts in the field of homeopathic interventions and as such qualified to give advice on this matter, but for non-medical homeopaths, matters of conventional medication usage falls outside their expertise and hence it would be unethical (and in contravention of their professional codes of conduct) to give advice on such usage.

Invariably, such issues raise concerns about respecting the autonomy of the patient. The right of the individual to self-select the treatments that they believe are right for them is highly respected and consequently, a homeopath may find him/herself in the position of trying to treat someone who is concurrently taking a combination of many conventional medications or, on the other hand, trying to treat a patient who wishes to stop conventional medication altogether.

Trying to respect the patient's wishes, while at the same time avoiding harm, can be challenging for any healthcare provider. Decisions about when it is appropriate and ethical to overrule autonomy in favour of wellbeing are complex and involve many factors; over-protection can infringe upon patient choice invoking paternalism, while under-protection can lead to harm. Advice about the best ways in which these two factors are kept in balance varies greatly but there is general agreement that, as far as possible, individuals must be informed about the likely benefits and risks of using any form of treatment, whether T&CM or conventional (Bruce and Robyn 2008).

Informed choice involves the exchange and understanding of relevant information so that a knowledgeable, reasoned and unpressured decision can be made (Gilmour et al. 2011b). However, for most forms of T&CM, including homeopathy, the shortage of data regarding efficacy and safety inevitably affect an individual's ability to make an informed choice about treatment options. If there is a paucity of research evidence available to support the use of certain treatments, then individuals need to be informed that this is the case. The reliability of claims that are made for some T&CM products and services have been seriously questioned (Ernst 2008), especially those advertised via the internet. The onus falls upon the T&CM community to improve the ways in which they inform people about products, practices and practitioners and to avoid unsubstantiated claims.

These challenges are not confined to the T&CM world; in conventional healthcare, practitioners can find themselves in comparable situations, but they normally have networks and support systems in place for themselves and their patients which are designed to support in such circumstances. Practitioners (of any kind) who work alone with people who have severe and enduring health problems, without support systems in place for themselves or their patient, put the safety of their patients at risk because they cannot possibly be available at all times or deal with every eventuality.

One possible solution to this challenge is that the practice of homeopathy is confined to conventional medical specialists, as it is in some European countries. If all practitioners are working within the conventional healthcare system there is less likelihood that diagnoses of serious illness will be missed, or that practitioners will be working alone. Another solution would be for the practice of homeopathy to be brought under statutory regulation so that homeopathic practitioners are regulated in the same way as those in conventional healthcare. However, neither of these two options is likely to happen in the UK, at least not in the short term.

If health is indeed, 'the ability to adapt and to self-manage' when facing physical, mental, and social challenges' (Huber et al. 2011), then consideration of what is needed for this to happen is warranted. For instance, for a person with mental health problems there may be many things that help them to adapt and manage; there could be a range of therapies, medications, lifestyle choices, social support systems and so on, that help. This might include homeopathy or conventional medicine or both.

The need for plurality in approach to health care is inherent within the current drive towards the provision of integrated care in the UK, which is an attempt to join up medical services offered by the NHS and social services offered by other bodies (Gov.uk 2014). Systems for integration are being put in place but they do not, as yet, include unregulated forms of T&CM. A consequence of this omission is that patients will continue to be at risk when they seek help from practitioners who work entirely on their own. For this situation to improve, there needs to be a willingness on both sides, from practitioners of T&CM and conventional medicine, to work together towards a common goal and for the good of the patient.

The preference for, and use of, one type of medicine over another is not in itself an ethical problem; the danger comes from overreliance on any one type of healthcare to the exclusion of all others. Many of the potential risks of T&CM stem from the notion that a particular modality can be *the* alternative, that it is the best and the only thing necessary for cure. Indeed, the use of the term 'alternative medicine' serves to promote the idea of one alternative, in this case an alternative to conventional medicine, as something to be used *instead of* conventional medicine. The potential for harm could be reduced through the eradication of the 'either/or' perspective. A truly patient-centred solution might be to treat all forms of healthcare as *complementary*, such that each is viewed as having its own strengths and weaknesses, and no one type of medicine is the best for all people in all situations.

The *WHO traditional medicine strategy 2014–2023* was developed to help countries determine how they can promote health and protect consumers who wish to use T&CM. In their strategy, the WHO specify that individual Member States will need to develop national approaches that reflect their specific needs in dealing with the most popular forms of T&CM practised in their country. Essential for this process, the WHO suggests, is that T&CMs are properly integrated into national health systems to enable consumers to have a wider choice when they wish to use such services (WHO 2013), a vision that is, as yet, far removed from the systems in most countries. Appropriate integration of healthcare systems has the potential to offer the best of both worlds. As suggested by Dr. Margaret Chan (2008), the Director General of WHO,

> The two systems of traditional and Western medicine need not clash. Within the context of primary health care, they can blend together in a beneficial harmony, using the best features of each system, and compensating for certain weaknesses in each. This is not something that will happen all by itself. Deliberate policy decisions have to be made. But it can be done successfully.

4.7.2 Recommendations for the Avoidance of Adverse Events from Homeopathy

The following recommendations apply to the homeopathic treatment of people by homeopathic practitioners who are not also qualified in conventional medicine.

For people using homeopathy the risk of experiencing AEs can be minimised if they:

- Consult with a conventional medical practitioner for diagnosis of their complaint before treatment, as most homeopaths are not qualified to undertake clinical diagnoses.
- Are aware that anyone in the UK (and many other countries) can legally call him/herself a homeopath and practise homeopathy, even if they have no training or experience, so it is important to ensure that their practitioner is registered with a professional body.
- Consult with a conventional practitioner on all matters concerning the use of conventional medications.

For practitioners, the risk of harm to their patients can be minimised if they:

- Always seek to work within their bounds of competence.[7]
- Are prepared to seek advice from their professional body and take challenging issues to supervision.

---

[7]When trying to assess competence the following are all important:

- Does the patient have support networks in place?
- Are other health professionals actively involved?
- Does the practitioner feel confident and comfortable with this case?
- What else needs to be put in place?
- Is regular supervision arranged?
- Who can be asked for help when needed?

- Never work alone.

  Ensure that patients with severe health problems have support networks in place before undertaking treatment.

  Ensure that both practitioner and patient know whom to contact when they need support/help/advice or if there is a medical emergency.

- Understand that they must take action if a patient poses a risk to themselves or others.

  Because of the potential for harm, patient confidentiality may be breeched in these situations.

- Never give advice about conventional medications.

  Because of the potential for harm, the patient should be clearly advised that medication withdrawal or reduction be undertaken in consultation with the person who has prescribed the medication.

References

Abdualmjid RJ, Sergi C (2013) Hepatotoxic botanicals—an evidence-based systematic review. J Pharm Pharm Sci 16(3):376–404

Bhagavathula AS, Elnour AA, Shehab A (2016) Pharmacovigilance on sexual enhancing herbal supplements. Saudi Pharm J 24(1):115–118

Bruce NW, Robyn AR (2008) Informed consent: good medicine, dangerous side effects. Camb Q Healthc Ethics 17(1):66–74

Carnes D, Mars TS, Mullinger B, Froud R, Underwood M (2010) Adverse events and manual therapy: a systematic review. Man Ther 15(4):355–363

Cawte J (1985) Psychoactive substances of the South Seas: Betel, Kava and Pituri. Aust NZ J Psychiatry 19(1):83–87

Chancellor A (2007) I'm becoming a less confident driver—especially in a country where the Vatican write the highway code. The Guardian. http://www.guardian.co.uk/commentisfree/2007/aug/24/comment.alexanderchancellor. Accessed 4 May 2018

Chatfield K, Duxbury J (2010) Practical and ethical implications for homeopaths in mental healthcare. In: Johannes CK, van der Zees H (eds) Homeopathy and mental healthcare: integrative practice, principles and research. Homeolinks, Holland

Chen S, Pang X, Song J, Shi L, Yao H, Han J, Leon C (2014) A renaissance in herbal medicine identification: from morphology to DNA. Biotechnol Adv 32(7):1237–1244

Connor KM, Payne V, Davidson JR (2006) Kava in generalized anxiety disorder: three placebo-controlled trials. Int Clin Psychopharmacol 21(5):249–253

Curtis P, Gaylord S (2005) Safety issues in the interaction of conventional, complementary, and alternative health care. Complement Health Pract Rev 10(3):31

Editorial (2005) The end of homeopathy. Lancet 366:690

Elliott R, Camacho E, Campbell F, Jankovic D, Martyn St James M, Kaltenthaler E, Wong R, Sculpher M, Faria R (2018) Prevalence and economic burden of medication errors in the NHS in England. Rapid evidence synthesis and economic analysis of the prevalence and burden of medication error in the UK. Policy Research Unit in Economic Evaluation of Health & Care Interventions. http://www.eepru.org.uk/wp-content/uploads/2018/02/eepru-report-medication-error-feb-2018.pdf. Accessed 5 May 2018

Ernst E (2003) Herbal medicines put into context: their use entails risks, but probably fewer than with synthetic drugs. BMJ 327(7420):881

Ernst E (2008) How the public is being misled about complementary/alternative medicine. J R Soc Med 101(11):528–530

Ernst E (2009) Harmless homeopathy? Int J Clin Rheumatol 4(1):7–10

Ernst E, Smith K (2018) More harm than good?: the moral maze of complementary and alternative medicine. Springer, Basel

European Committee for Homeopathy (2018) Regulatory status. European Council for Homeopathy. http://www.homeopathyeurope.org/regulatory-status. Accessed 15 Mar 2018

Farah M (2011) Classification and monitoring safety in herbal medicine. WHO-UMC, Sweden

Farah MH, Edwards R, Lindquist M, Leon C, Shaw D (2000) International monitoring of adverse health effects associated with herbal medicines. Pharmacoepidemiol Drug Saf 9(2):105–112

Fisher P (2006) Homeopathy and the Lancet. Evid-Based Complement Altern Med 3(1):145

Freckelton I (2012) Death by homeopathy: issues for civil, criminal and coronial law and for health service policy. J Law Med 19(3):454–478

Future Market Insights (2017) Herbal medicines product market. https://www.futuremarketinsights.com/reports/herbal-medicinal-products-market. Accessed 3 Mar 2018

Gale NK, McHale JV (2015) Routledge handbook of complementary and alternative medicine: perspectives from social science and law. Routledge, New York

Gallo E, Pugi A, Lucenteforte E, Maggini V, Gori L, Mugelli A, Firenzuoli F, Vannacci A (2014) Pharmacovigilance of herb-drug interactions among preoperative patients. Altern Ther Health Med 20(2):13–17

Garrouste-Orgeas M, Philippart F, Bruel C, Max A, Lau N, Misset B (2012) Overview of medical errors and adverse events. Ann Intensive Care 2(1):2

Gilard V, Balayssac S, Tinaugus A, Martins N, Martino R, Malet-Martino M (2015) Detection, identification and quantification by 1H NMR of adulterants in 150 herbal dietary supplements marketed for improving sexual performance. J Pharm Biomed Anal 102:476–493

Gilmour J, Harrison C, Asadi L, Cohen MH, Vohra S (2011a) Complementary and alternative medicine practitioners' standard of care: responsibilities to patients and parents. Pediatrics 128(4):S200–S205

Gilmour J, Harrison C, Asadi L, Cohen MH, Vohra S (2011b) Informed consent: advising patients and parents about complementary and alternative medicine therapies. Pediatrics 128(4):S187–S192

Gov.uk (2014) Enabling integrated care in the NHS. UK Government. https://www.gov.uk/enabling-integrated-care-in-the-nhs. Accessed 2 May 2018

Grollman AP, Shibutani S, Moriya M, Miller F, Wu L, Moll U, Suzuki N, Fernandes A, Rosenquist T, Medverec Z (2007) Aristolochic acid and the etiology of endemic (Balkan) nephropathy. Proc Natl Acad Sci 104(29):12129–12134

Heaton CA (1994) The chemical industry. Springer Science & Business Media, Berlin

Herbal Medicines Advisory Committee (2017) Annual Report 2016. https://www.gov.uk/government/groups/herbal-medicines-advisory-committee#annual-report. Accessed 5 Mar 2018

Hope A (2010) Coroner's report: inquest into the death of Penelope Dingle. http://www.homeowatch.org/news/dingle_finding.pdf. Accessed 5 Apr 2018

Huber M, Knottnerus JA, Green L, van der Horst H, Jadad AR, Kromhout D, Leonard B, Lorig K, Loureiro MI, van der Meer JWM, Schnabel P, Smith R, van Weel C, Smid H (2011) How should we define health? British Med J 343:d4163

Hunt KJ, Ernst E (2010) No obligation to report adverse effects in British complementary and alternative medicine: evidence for double standards. Qual Saf Health Care 19(1):79

Kim H, Hughes PJ, Hawes EM (2014) Adverse events associated with metal contamination of traditional chinese medicines in Korea: a clinical review. Yonsei Med J 55(5):1177–1186

Kohn LT, Corrigan JM, Donaldson MS (2000) To err is human: building a safer health system. National Academies Press, Washington

Le J (2014) BCPS-ID "Pharmacokinetics". Merck Manual for Healthcare Professionals. http://www.merckmanuals.com/professional Accessed 16 Apr 2018

Leape LL, Berwick DM (2005) Five years after To Err Is Human: What have we learned? J Am Med Assoc 293(19):2384–2390

Lim A, Cranswick N, South M (2011) Adverse events associated with the use of complementary and alternative medicine in children. Arch Dis Child 96(3):297–300

Linde K, Ramirez G, Mulrow CD, Pauls A, Weidenhammer W, Melchart D (1996) St John's wort for depression—an overview and meta-analysis of randomised clinical trials. BMJ 313(7052):253–258

Luteijn JM, White BC, Gunnlaugsdóttir H, Holm F, Kalogeras N, Leino O, Magnússon SH, Odekerken G, Pohjola MV, Tijhuis MJ, Tuomisto JT, Ueland Ø, McCarron PA, Verhagen H (2012) State of the art in benefit–risk analysis: medicines. Food Chem Toxicol 50(1):26–32

MacPherson H, Thomas K, Walters S, Fitter M (2001) The York acupuncture safety study: prospective survey of 34,000 treatments by traditional acupuncturists. BMJ 323(7311):486

Medicine and Healthcare products Regulatory Agency (2014) Banned and restricted herbal ingredients. https://www.gov.uk/government/publications/list-of-banned-or-restricted-herbal-ingredients-for-medicinal-use/banned-and-restricted-herbal-ingredients. Accessed 1 Mar 2018

Moncrieff J, Cohen D, Porter S (2013) The psychoactive effects of psychiatric medication: the elephant in the room. J Psychoactive Drugs 45(5):409–415

Nissen N, Weidenhammer W, Schunder-Tatzber S, Johannessen H (2013) Public health ethics for complementary and alternative medicine. Eur J Integr Med 5(1):62

Okoronkwo I, Onyia-Pat J-L, Okpala P, Agbo M-A, Ndu A (2014) Patterns of complementary and alternative medicine use, perceived benefits, and adverse effects among adult users in Enugu Urban. Evid-Based Complement Altern Med, Southeast Nigeria. https://doi.org/10.1155/2014/239372

Pirotta M, Willis K, Carter M, Forsdike K, Newton D, Gunn J (2014) 'Less like a drug than a drug': the use of St John's wort among people who self-identify as having depression and/or anxiety symptoms. Complement Ther Med 22(5):870

Posadzki P, Watson L, Ernst E (2013a) Adverse effects of herbal medicines: an overview of systematic reviews. Clin Med 13(1):7–12

Posadzki P, Watson L, Ernst E (2013b) Contamination and adulteration of herbal medicinal products (HMPs): an overview of systematic reviews. Eur J Clin Pharmacol 69(3):295–307

Posadzki P, Watson L, Ernst E (2013c) Herb-drug interactions: an overview of systematic reviews. Br J Clin Pharmacol 75(3):603–618

Rajendran D, Bright P, Bettles S, Carnes D, Mullinger B (2012) What puts the adverse in 'adverse events'? Patients' perceptions of post-treatment experiences in osteopathy—a qualitative study using focus groups. Manual Therapy 17(4):305–311

Schulze J, Raasch W, Siegers C-P (2003) Toxicity of kava pyrones, drug safety and precautions—a case study. Phytomedicine 10:68–73

Science and Technology Committee (2010) House of Commons Science and Technology Committee—Fourth Report. Evidence Check 2: Homeopathy. http://www.publications.parliament.uk/pa/cm200910/cmselect/cmsctech/45/4502.htm Accessed 10 Feb 2018

Scott S, Thompson J (2014) Adverse drug reactions. Anaesth Intensiv Care Med 15(5):245–249

Sehon S, Stanley D (2010) Evidence and simplicity: Why we should reject homeopathy. J Eval Clin Pract 16(2):276–281

Shang A, Huwiler-Muntener K, Nartey L, Juni P, Dörig S, Sterne JAC, Pewsner D, Egger M (2005) Are the clinical effects of homoeopathy placebo effects? Comparative study of placebo-controlled trials of homoeopathy and allopathy. Lancet 366(9487):726–732

Smith SD, Stephens AM, Werren JC, Fischer GO (2013) Treatment failure in atopic dermatitis as a result of parental health belief. Med J Aust 199(7):467–469

Society of Homeopaths (2018) Code of Ethics. Society of Homeopaths. https://homeopathy-soh.org/resources/code-of-ethics/soh-codeethics-2015_r1/. Accessed 1 May 18

Spence DS, Thompson EA, Barron SJ (2005) Homeopathic treatment for chronic disease: a 6-year, university-hospital outpatient observational study. J Altern Complement Med 11(5):793–798

Stickel F, Shouval D (2015) Hepatotoxicity of herbal and dietary supplements: an update. Arch Toxicol 89(6):851–865

Stone J (2000) Ethical issues in complementary and alternative medicine. Complement Ther Med 8(3):207–213

Sultana J, Cutroneo P, Trifirò G (2013) Clinical and economic burden of adverse drug reactions. J Pharmacol Pharmacother 4(11):S73

Supreme Court of India (1996) Poonam Verma vs. Ashwin Patel. http://www.indiankanoon.org/doc/611474. Accessed 13 Nov 2017

Teschke R, Schulze J, Schwarzenboeck A, Eickhoff A, Frenzel C (2013) Herbal hepatotoxicity: suspected cases assessed for alternative causes. Eur J Gastroenterol Hepatol 25(9):1093–1098

Thompson R, Ruch W, Hasenöhrl RU (2004) Enhanced cognitive performance and cheerful mood by standardized extracts of Piper methysticum (Kava-kava). Hum Psychopharmacol: Clin Exp 19(4):243–250

Tsang C, Bottle A, Majeed A, Aylin P (2013) Adverse events recorded in English primary care: observational study using the General Practice Research Database. British J General Pract 63(613): e534-e542

Tweed V (2015) Herb-drug interactions lack evidence. Better Nutr 77(6):12

Uppsalla Monitoring Centre (2013) Glossary of Pharmacovigilance terms. Available at: https://www.whoumc.org/global-pharmacovigilance/global-pharmacovigilance/glossary/

Uppsala Monitoring Centre (2018) What is VigiBase? https://www.who-umc.org/vigibase/vigibase/. Accessed 27 Apr 18

Vanherweghem J-L (1998) Misuse of herbal remedies: the case of an outbreak of terminal renal failure in Belgium. J Altern Complement Med 4(1):9–13

Vickers A, Zollman C, Lee R (2001) Herbal medicine. West J Med 175(2):125

Vithoulkas G (1980) The science of homeopathy. Grove Press, New York

WebMD (2018) What is penicillin allergy? https://www.webmd.com/allergies/penicillin-allergy#1. Accessed 03 Apr 2018

Weich S, Brugha T, King M, McManus S, Bebbington P, Jenkins R, Cooper C, McBride O, Stewart-Brown S (2011) Mental well-being and mental illness: findings from the Adult Psychiatric Morbidity Survey for England 2007. Br J Psychiatry 199(1):23–28

Werner SM, Soghomonyan S (2014) Patient safety and the widespread use of herbs and supplements. Front Pharmacol 5:1–2

White A, Hayhoe S, Hart A, Ernst E (2001) Adverse events following acupuncture: prospective survey of 32 000 consultations with doctors and physiotherapists. BMJ 323(7311):485–486

Wooltorton E (2002) Herbal kava: reports of liver toxicity. Can Med Assoc J 166(6):777

World Health Organization (1972) International drug monitoring: the role of national centres. Report of a WHO meeting, Geneva: World Health Organization

World Health Organization (2004) Pharmacovigilance: ensuring the safe use of medicines. World Health Organization, Geneva

World Health Organization (2013) WHO traditional medicine strategy: 2014–2023. World Health Organization, Hong Kong SAR

World Health Organization (2014) 10 facts on patient safety. http://www.who.int/features/factfiles/patient_safety/en/. Accessed 10 Mar 2018

World Health Organization (2018) Patient safety. http://www.who.int/patientsafety/en/. Accessed 10 Mar 2018

Chapter 5
Ethical Challenges for Animals from Traditional and Complementary Medicine

Abstract Humans use and abuse animals for many purposes, including the production and testing of traditional and complementary medicine (T&CM) products. While some animals may benefit from T&CM interventions, many more suffer harm. In both animal experimentation and for the production of T&CM products, animals can be exposed to stress, pain, artificially induced diseases and/or ultimately killed. However, the use of animals in T&CM products (for example, oil extracted from the blubber of the River Dolphin or the intestines of a porcupine) is more ethically challenging than the use of animals in T&CM research. First, research is better regulated. Second, animals used in research are often bred for purpose whereas most of the animals used for T&CM products are taken from the wild in an unregulated manner and many are endangered species such as rhinos, and tigers. Ethical challenges for the use of animals in T&CM products are analysed together with potential for adoption of the same ethical principles that govern animal experimentation (replacement, reduction, refinement).

Keywords Traditional medicine · Animals · Animal welfare · Animal ethics Three R's · Zootherapy

5.1 Animals in the Ethical Matrix

For the application of the ethical matrix, Mepham asks us to consider the ethical dimensions of particular issues in accordance with the three principles of wellbeing, autonomy and justice (Mepham 1995). However, contentious debate about the applicability of autonomy and justice to non-human animals is enduring and even amongst those who support such an application there is no agreement as to how the principles should be interpreted. The inclusion of these principles in the analysis of ethical challenges for animals would be similarly contentious and open to broad-based criticism. Hence, for animals, a strictly consequentialist approach is adopted here.

My analysis of ethical concerns begins with an overview of the use of animals in conventional medical products in order to provide background context.

5.2 The Use of Animals in Conventional Medicine

Certain conventional medications either contain animal products or are derived from animal sources. The Queensland Department of Health (2013) list pharmaceutical products known to be of animal origin including fifteen drugs derived from pigs, fifteen from cows, ten from mice, ten from chicken eggs, thirty-six from Chinese hamster ovary cells, thirteen from horses and a further twenty from a range of other sources, such as chondroitin that is derived from bovine or shark cartilage and Digibind that is derived from sheep. Production of some of these drugs involves the killing of animals. For example, heparin, an injectable anticoagulant, is derived from the tissues of slaughtered animals, mainly pigs. Attempts have been made to produce heparin synthetically but, as yet, this approach does not offer a viable alternative approach to produce a drug that is in high worldwide demand (Pavão and Mourão 2012).

Many of the drugs that are derived from animal products do not involve the slaughtering of animals, such as herceptin. Herceptin inhibits the proliferation of human tumor cells under certain conditions and is commonly used in the treatment of patients with metastatic breast cancer. This drug is produced by a cell line[1] that is derived from the ovaries of a Chinese Hamster. These cells are the most commonly used mammalian sources for the industrial production of therapeutic proteins (Omasa et al. 2010).

One particular pharmaceutical drug of animal origin has received a considerable amount of criticism and is the subject of ongoing debate. The derivation of its name, Premarin, stems from its source in 'pregnant mares' urine', and is manufactured from conjugated oestrogens obtained from the urine of pregnant horses. It is taken by millions of women worldwide to treat the symptoms of menopause. Most of the world's Premarin is derived from the urine of approximately two thousand pregnant horses kept on twenty-six ranches in North Dakota and Canada (North American Equine Ranching Information Council 2015). Opponents of Premarin production claim that both the pregnant mares and their foals suffer greatly:

> For most of their 11-month pregnancies, the horses are confined to stalls so small that they cannot turn around or take more than a single step in any direction. The animals must wear rubber urine-collection bags at all times, which causes chafing and lesions, and their drinking water is limited so that their urine will yield more concentrated estrogen. Once the foals are born, the horses are impregnated again, and this cycle continues for about 12 years (People for the Ethical Treatment of Animals 2018).

---

[1]A cell line is a population of cells descended from a single cell and containing the same genetic makeup. The cells are reproduced in the laboratory.

To many horse-lovers, the nature of this industry is considered cruel and profoundly disturbing (Andersson 2004), but the North American Equine Information Council, representing *ranchers* involved in the collection of pregnant mares' urine, takes a different view. They point out that all ranchers must adhere to the '*Recommended Code of Practice for the Care and Handling of Horses in PMU Operations*', which specifies standards for nutrition, watering, exercise, and barn environment, among other requirements. Furthermore, they claim that equine ranching is the most highly regulated horse ranching activity in North America (North American Equine Ranching Information Council 2015).

As well as their use in pharmaceutical products, many animal tissues and their derivatives are used in medical devices. These materials can comprise a major part of the device, such as bovine/porcine heart valves or bone substitutes for use in dental or orthopaedic applications or they may be used as a product coating, as is the case with collagen or gelatine. Some biochemicals of animal origin are used in the device manufacturing process, such as stearates[2] from animal fats, or foetal bovine serum[3] (Silver 2012).

European regulations governing the production of animal-derived products and devices are largely concerned with human safety issues and risks to those that use them. For example, European Union Regulation 722/2012, which covers medical devices that use animal tissues, requires device manufacturers to ensure adequate risk management and controls to prevent spreading certain animal-borne diseases to human users of their products. However, manufacturers are also required to provide a justification for the use of animal tissues or derivatives, specifying animal species, tissues and sourcing and taking into account the clinical benefit, potential residual risk and suitable alternatives, such as lower risk tissues or synthetic alternatives (European Union 2012).

5.3 Animals, T&CM and Wellbeing

Animals can benefit from T&CM and many forms are used in the treatment of animals. For example, in the UK, the Association of British Veterinary Acupuncturists, describe how acupuncture is often used to treat pain in cats, dogs and horses (ABVA 2018) and the British Association of Veterinary Herbalists tell us that herbal supplements can be used for treatment of a wide range of common conditions (BAVH 2018). In the Western world, interest in the use of these types of T&CM for animals is increasing. Just as some people choose to use T&CM for their own treatment, they also choose it for the treatment of their animals (Bergenstrahle and Nielsen 2015). Globally, in regions where people rely upon T&CM products and services for themselves, this may also be the only treatment option available for animals.

---

[2]Stearates are commonly used in the production of medicines as a lubricant. It prevents the ingredients from sticking to the manufacturing equipment.

[3]Foetal bovine serum is the most widely used growth supplement for cell cultures.

While there is a potential for benefit from the use of T&CM, there is also the potential for immense harm to animals. The traditional practice of using animals or animal-derived products for medicinal purposes has been termed 'zootherapy' (Costa-Neto 1999) and appears to have deep historical origins in many different countries around the world. Many T&CM products are derived from animals and investigations have been undertaken in a number of regions where zootherapy is commonplace to document usage. These investigations most often take the form of interviews, discussions or focus groups with traditional healers to elicit information that is customarily passed on through an oral tradition from generation to generation. The results from these ethnographic studies are helping to build a picture of global usage: which animals are used, how the animals are obtained, what parts of the animals are used, and for what ailments.

From the combined findings of different researchers, together with information already in the literature, it is possible to get some indication of the species of animals that are used in T&CM products in certain regions. For instance, Alves and Alves (2011) found 584 animal species were reportedly used in traditional medicines in Latin America; 354 of these in Brazil alone (Alves et al. 2013). In India, where there is an immense biodiversity of fauna, accounting for 10% of the reported biological species on the planet (Kim and Song 2013), there is documented evidence for at least 351 medicinal uses of animal and animal parts (Sajem Betlu 2013). In Bhutan, Yeshi et al. (2017) identified 45 animal species are used in Bhutanese Rigpa medicine, of which 26 were mammals (including bear, deer, sheep, goat, yak, and elephant). These numbers are probably an underestimate of actual usage, since it is unlikely that all uses have been documented.

In the case of China, where the use of animals is long established in Traditional Chinese Medicine, no documentation of an ethnographic nature on the use of animals could be found.[4] However, numbers have been postulated using information drawn from ancient and current textbooks of Traditional Chinese Medicine and it is estimated that as many as 1500 animal species are used in T&CM (Still 2003).

Traditionally, numerous animal parts have been claimed to have medicinal properties and the use of medicines obtained from animals is suggested as traditional cures for a wide variety of ailments. Some treatments involve use of the whole animal, while for others, specific parts are used including flesh, fat, blood, milk, eggs, bile, excrement, bone, horns, tusks, teeth, nails, feathers, fur, penis and all internal organs. The range of ailments that are treated with animal products is very broad encompassing first aid (such as injuries, burns, poisoning), acute ailments (such as fevers and infections), as well as long-term, chronic conditions (such as jaundice, diabetes and cancers).

Some of the relevant animal species are critically endangered. For instance, one of the most highly prized animals is the tiger, sought for the use of various body parts, especially the bones, but also the hair, teeth, skin and many other organs (Still 2003). Tiger parts are used to treat a number of disorders; examples include the use of the eyeballs for epilepsy; the whiskers for toothache; the urine for rheumatism;

---

[4]The search precluded any publications not written in English.

the bones for joint problems and hemiplegia; the brain for laziness and pimples; and the tail for skin diseases (Athiyaman 2008).

The illegal trade of animal products for medicinal use has contributed to a decrease in numbers of many animal populations (Rangarajan 2005), clearly illustrated by the case of the rhino. Rhinos, once common throughout Eurasia and Africa, are now found almost exclusively in national parks and reserves. Poaching is a threat in all rhino habitats and, as the numbers of rhinos declines, the intermittent availability of rhino horn drives the price higher, thereby increasing the incentive to poach (Save the Rhino 2018). Two species of rhino in Asia (the Javan and Sumatran rhinos) are classified as 'critically endangered' (WWF 2018). The decline in rhino numbers is undoubtedly linked to its use in T&CM. Virtually every part of the rhino is used: the horn for alleviating fever, the skin for treating skin disease, the penis as an aphrodisiac, the bone to treat bone disorders and the blood as a tonic for women who are suffering from menstrual problems (Still 2003). Smaller Asian rhino horns are more highly prized than African rhino horns because it is believed they will be more potent but animal studies undertaken in the UK and South Africa found no pharmacological basis for the claimed effects of rhino horn, or any other animal horns (Abraham 2014).

The use of non-human primates in T&CM is also widespread and has been recorded in fifty-one countries, mainly in Latin America, Africa and Asia. According to Alves et al. (2010), the trade in these products is encouraging the commercial hunting of primates, and threatening them on a scale they have never faced before. Of the 101 species of primates recorded in their review, Alves et al. found that twelve species were classified as 'Critically Endangered', twenty-three as 'Endangered', twenty-two as 'Vulnerable' and seven as 'Near Threatened'.

Another large mammal that is in demand for its supposed medicinal properties is the bear. Bear bile has been prescribed in Traditional Chinese Medicine for centuries and it is used to treat a range of complaints including inflammation, bacterial infections and pain (Still 2003). Widespread illegal killing of bears is contributing to declining numbers, especially in Southeast Asia and China (International Union for Conservation of Nature 2011). In the 1980s, with numbers of wild bears in decline, the practise of bear farming was established and today, there are estimated to be 10,000 bears on Chinese bear farms alone (Bear Conservation 2018). Many have expressed concerns about their welfare as the bears are commonly kept for years in tight cages; their teeth are broken, their claws are pulled out and the bile collection process can cause severe pain through use of a catheter or implant into the bear's gall bladder (Yibin et al. 2009). When the bears cannot produce sufficient bile they are often left to die of starvation (Kikuchi 2012). The most important component of bear bile is ursodeoxycholic acid and this has been shown to be effective against many complaints including liver and gall bladder ailments (Challem 2007). However, this chemical can be synthesised artificially (Still 2003). Indeed, the synthesised version is commonly used for the treatment of cystic fibrosis, gallstones and biliary cirrhosis (Electronic Medicines Compendium 2018).

In Latin America, where zootherapies form an integral part of the local culture, many of the animal-derived medicines include threatened species (Costa-Neto 1999).

For instance, of the 354 animal species known to be used in Brazil, 21% are on one or more lists of endangered species (Alves and Albuquerque 2013). Most of the medicinal animals used in Brazil are wild and although Brazilian legislation forbids commercial use of wild fauna, medicinal products and derivatives made from animals, including species that are on the list of endangered species, are commonly traded in Brazilian markets (Alves 2009). This trade occurs illicitly and without due monitoring by competent environmental agencies (Ferreira et al. 2013); in Brazil, the trade in medicinal animals is described as 'a well organised practice' (Ferreira et al. 2015).

Given this situation, the desire to investigate alternatives to the use of animals in medicine appears to be growing and suggestions have been made for the substitution of the use of threatened animal species with medicinal plants or domestic animals (Luo et al. 2011). This is a complex process because the use of animals for medicinal purposes can be related to a host of factors including biological, cultural and socioeconomic aspects (Rastogi and Kaphle 2011).

5.4 Animal Ethics

Consideration of exactly how the interests of animals should be represented, awarded value and weighed against the competing interests of humans could easily fill several books. Hence, the following brief ethical analysis seeks only to expose specific points of contention and reveal useful recommendations for T&CM practice.

The idea that animals should be awarded moral consideration was greatly advanced in utilitarian ethics; as the founder of modern utilitarianism, Jeremy Bentham, famously quoted: 'The question is not, Can they *reason*? nor, Can they *talk*? but, Can they *suffer*?' (Bentham 1823). Bentham was not against the use of animals for food or experimentation but he expounded the view that suffering should be avoided and that animals should only be used if there was a realistic potential for good (of humanity). In other words, the potential for good to humans would need to outweigh the potential for harm to animals. Sensitivity towards the suffering of animals was also advanced by the work of Charles Darwin who starkly challenged the notion that humans were created by God in his own image: 'Man in his arrogance thinks himself a great work worthy the interposition of a deity. More humble, and I believe truer, to consider him created from animals' (Darwin 1838). Darwin was resolute in his opinion that many animals, just like humans, are sentient beings:

> The lower animals, like man, manifestly feel pleasure and pain, happiness and misery. Happiness is never better exhibited than by young animals, such as puppies, kittens, lambs, &c., when playing together, like our own children (Darwin 1871).

In more recent years, two ground-breaking and highly influential books concerning animal ethics have dominated discussions in this area; *Animal Liberation* by Peter Singer (1975) and *The case for Animal Rights* by Tom Regan (1987). Regan claims that sentient beings who are able to see themselves as 'subjects of life' have 'inherent value' which awards them defensible moral rights and that this in turn implicates *prima facie* duties for human beings towards animals. Singer, on the other hand, argues for utilitarian based animal ethics grounded in the notion of 'equal consideration of interests'. Equal consideration of interests requires us to give equal weight to similar interests, regardless of species. Preference for the interests of one species over another constitutes 'speciesism', a prejudice that is held no more justifiable than sexism or racism. A key point in Singer's argument is that equal consideration of interests is necessary for *sentient* beings that are capable of feeling pain and suffering.

Given that I am taking a consequentialist approach to the analysis of ethical issues associated with animals and T&CM, coupled with the requirement to place myself in the 'shoes of animals', it seems fitting to follow Singer's line of reasoning in this analysis. Maintenance of sensitivity towards speciesism helps to avoid anthropocentric bias in the weighing of the potential harms and benefits necessary for a consideration of animal wellbeing.

5.5 Ethical Analysis: The Use of Animals in T&CM Products

There is obvious harm to animals that are used in T&CM medicinal products. Many wild animals are hunted and killed and excessive or uncontrolled hunting has led to the extinction or near extinction of some species. Others, who are not killed, may suffer great distress and pain, such as the bears that are kept on bear farms for their bile.

Medicinal benefit from animal-derived products (for humans or animals) is far from certain, even for the most expensive and widely used products. Most have never been tested for efficacy; some that have been tested show no benefit (like rhino horn). Hence, in the main, animals are being harmed for the manufacture of products of unproven value. There could be other ways, aside from medicinal effects, in which humans benefit from the use of animal-derived T&CM. Zootherapy may be an important and integral part of a traditional healthcare system with cultural and economic significance. The values of animal-based medicine are important in tribal culture. They are easily available resources for people with limited access to other types of medicine (Das 2015). However, the killing and/or suffering of animals to make T&CM products for the benefit of humans is inherently speciesist. If we award the bear equal consideration of interests to the human recipient of bear bile, then it is clear that the painful extraction of bear bile is not justifiable because the bear's interests are sacrificed to serve human interests. This is especially difficult to justify when there is a readily available synthesised version of the active ingredient of bear bile that could be used as an alternative. The acquisition of bear bile causes pain and suffering. The manufacture of a synthesised equivalent does not.

From a utilitarian perspective, it can be argued that potential benefits to a large number of beings might justify the humane killing of one or a few beings. For example, the preparation of the homeopathic medication, *Apis melifica*, requires the killing of one honeybee, but the resultant medication can be used to treat thousands of humans and animals. Not many would argue that this is unethical. The principle of equal consideration of interests does not apply if a being is incapable of suffering or feeling pain or happiness. Bees are not considered to be sentient creatures and hence, in this case, there is no competing interest to take it into account.

However, the killing of a rhino, a member of an endangered species, for treatments that are of dubious medicinal value, is hard to justify, even if the resultant products are distributed to many. There is a quantitative and qualitative difference between the killing of a bee and the killing of a rhino.

According to Singer, we must take care when we compare the interests of different species. In some situations, a member of one species will suffer more than a member of another species. In this case we should still apply the principle of equal consideration of interests, but we must give priority to relieving the greater suffering (Singer 1993). This seems to suggest that it is ethically acceptable to use some animals in T&CM products if the suffering of the used animal(s) is less than the suffering that the potential recipients would experience without this medication. However, the weighing of relative degrees of suffering between individuals is extremely difficult, and even more so when they are of different species.

It is clear that the use of animals in T&CM products is rarely (if ever) undertaken in the interests of animals. Most animal-derived T&CM products are intended for human use and human benefit. Here, the ethical matrix exposes a conflict between the wellbeing of animals and the wellbeing of humans. Given that there may be some circumstances in which the use of animals in T&CM products is justifiable (when interests have been awarded equal consideration) and there is no easy way of weighing relative degrees of suffering, it is helpful to consider application of existing principles that have been developed to protect the welfare of animals, just as we have for animal experimentation.

5.5.1   The Regulation of Animal Use in Medical Research

Most existing animal research policies globally are concerned with animal welfare and are underpinned by the notion of the 'three Rs': replacement, reduction, and refinement, first proposed by William Russell and Rex Burch in 1959 (Russell et al. 1959). *Replacement* refers to the idea that, whenever possible, the use of animals should be replaced with an alternative method that does not employ sentient creatures. When replacement is not possible, *reduction* requires that the lowest number of animals is used that are needed for obtaining meaningful results, and *refinement* refers to any factors that can decrease the incidence or severity of inhumane procedures for all animals that are used.

In other words, sentient animals should only ever be used in experimentation when there is no other means of addressing a (justifiable[5]) research problem. If animals are to be used, care must be taken to use enough animals to ensure significance of the results and no more. Additionally, these animals must be treated in a humane manner.

The three Rs provide the foundation for European policy, Directive 2010/63/EU on the protection of animals used for scientific purposes, which took full effect on 1 January 2013. The Directive stipulates measures that must be taken to replace, reduce and refine the use of animals and regulates the use of animals through systematic project evaluation requiring the assessment of pain, suffering, distress and lasting harm caused to the animals (European Commission 2018).

Globally, while there are variations in practical application, regulations in individual countries are generally based on these same principles (Guillén and Vergara 2018). Additionally, there are international efforts to harmonise standards and develop global frameworks. For example, the International Council for Laboratory Animal Science (ICLAS) and the Council for International Organizations of Medical Sciences (CIOMS) have produced 'International Guiding Principles for Biomedical Research Involving Animals' (CIOMS and ICLAS 2012). The ICLAS-CIOMS principles incorporate the Three Rs and are intended to serve as a framework of responsibility for all countries, including those with emerging research programmes.

Implementation of the three Rs has undoubtedly resulted in improvements to wellbeing in animal research globally. Furthermore, their broad adoption signals a growing awareness and acceptance of the need for global ethical standards regarding the use of animals for medical purposes. Against this backdrop, it seems quite reasonable to suggest that the three Rs should also be applied universally for ethical assessment of animal use in other human healthcare activities. There is no good reason why, if it is a requirement for the use of animals in experimentation, that it should not also be a requirement for the use of animals in the manufacture of healthcare products. The fact that this notion is readily accepted within animal research, but not even discussed in regard to use of animals in the production of T&CM medicinal products, is a clear example of double standards.

5.5.2 Application of the Three Rs for Animal Use in T&CM Products

Given the range of cultural sociological, political, economic and legal factors at play, it might seem idealistic to propose that the three Rs could be applied to the use of animals in T&CM. Humans have long exploited animal products for food, materials (leather/fur), labour, medicines and so on. Our culture and our experiences guide our attitudes towards animals (Alves et al. 2018) and animal products have a long history

---

[5]This also involves a weighing of harms and benefits; the potential benefits of the research must outweigh the potential for harm to the animals.

as important components of traditional medicines and rituals in numerous regions (Williams and Whiting 2016). Additionally, animals are commonly used in T&CM in regions where resources are poor and regulations are minimal. Nevertheless, the three Rs can be used as guiding principles for ethical assessment to ensure that the interests of animals are awarded due consideration.

Application of the three Rs to the use of animals in T&CM products would require consideration of the following factors.

Replacement: Can the animal product be replaced with any other medicinal product? Sentient animals must be replaced in all possible circumstances. Replacement of an animal-derived product with a plant-based product, conventional drug or even a product that is derived from non-sentient animals would reduce suffering.

While this brief analysis may give the impression that the use of animals in T&CM is ubiquitous, in fact, the vast majority of traditional treatments are of botanical origin (Williams and Whiting 2016). Furthermore, most conditions for which animal products are used, are also treated with plant based products. In Brazil, Nascimento et al. (2016) found that of a total of 237 different therapeutic targets, 150 were treated exclusively with plants, 19 exclusively with animals and 68 could be treated with either. Plants were used to treat a greater number of therapeutic targets than animals, and most targets treated with animals were also treated with plants. Additionally, an investigation of user preferences found no difference between preference for plant and animal medicines. The primary deciding factor appeared to be availability rather than preference, as both sources were considered equally effective.

The application of the principle of replacement does not preclude the use of animals in T&CM, it simply requires us to consider and use alternatives when possible. This is important because, for some, animals may provide their only means/access to a healthcare product. However, it appears that in many situations, there may be a pragmatic choice and in such cases, the choice of plant-based medicine is generally preferable.[6]

Reduction: How is the need for the T&CM medication being measured and what is the minimum amount required to meet the real need? In order to reduce the number of animals being used, the actual need for a product must be explicit and quantified.

The use of animals in T&CM can have devastating consequences that extend beyond the individual creature and affect an entire species. The rhino is a perfect example but the hunting of rhinos for T&CM does not generate ethical dilemmas. The hunting of rhinos is simply wrong, no debate is necessary. Rhinos are a protected species and the killing of rhinos for use in T&CM is illegal. Aside from this, tests have failed to demonstrate any therapeutic benefit from use of rhino horn and hence their use could never be justified. For most animal-derived T&CMs, there are no test results to help guide decisions about reduction. Knowledge about the use of animals in T&CM by different ethnic communities is generally passed orally from one generation to the next (Borah and Prasad 2017). Hence, for now, decisions about

---

[6]In certain situations, this decision might be complicated. Consider, for example, the choice between use of an animal by-product from local livestock that would otherwise be discarded, and an endangered plant species that only grows in a treacherous domain.

reduction can only be made in a generalised fashion with advice to use 'as little as possible to achieve the required effect'.

Refinement: What measures are in place to minimise the pain, suffering, distress or harm to the animals? In the cases where a real need has been demonstrated and there is no possibility of replacement, then suffering to the animal must be reduced and welfare maximised.

For individual animals, it is relatively easy to imagine what refinement would entail. Application of this principle would minimise suffering to the animal, preclude maltreatment and the use of extremely injurious practices like bear farming altogether. However, as with reduction, this principle may also be extended to include an assessment of the welfare of a community of animals or even the entire species. How much does harm to one, impact upon the welfare of others or the whole? This is particularly important to consider in cases where animals are taken from the wild, where they are not bred for purpose and their use may have damaging effect that extends beyond the individual.[7]

5.6 Recommendations for the Use of Animals in T&CM Products

The same ethical standards should be applied to both the use of animals in research and the use of animals in the manufacturing of T&CM products.

There may be differences in how the standards are applied but the principles should be the same. For example, the principle of reduction is intended to minimise the number of animals used. In research this would involve an understanding of methodological design and statistical analysis; in the manufacture of products this would involve an accurate assessment of the actual need for a product.

Ethical standards for the use of animals need to be globally agreed and applied consistently.

The current movement towards global standards for animal experimentation should be encouraged and developed further.

References

Abraham C (2014) Rhino horn is no medicine, New Sci 222 (2966): 27–27
Alves RRN (2009) Fauna used in popular medicine in Northeast Brazil. J Ethnobiol Ethnomed 5:11
Alves RRN, Albuquerque UP (2013) Animals as a source of drugs: bioprospecting and biodiversity conservation. In: Animals in traditional folk medicine. Springer, Berlin, pp 67–89

---

[7]For instance, killing of a nursing mother would obviously impact upon her infants. More broadly, killing certain animals, even if not endangered, may have broad impacts upon the sustainability of their community.

Alves RRN, Alves HN (2011) The faunal drugstore: animal-based remedies used in traditional medicines in Latin America. J Ethnobiol Ethnomed 7(1):9–51

Alves RRN, Santana GG, Rosa IL (2013) The role of animal-derived remedies as complementary medicine in Brazil. In: Animals in traditional folk Medicine. Springer, Berlin, pp 289–301

Alves RRN, Silva JS, da Silva Chaves L, Albuquerque UP (2018) Ethnozoology: an overview and current perspectives. In: Ethnozoology. Elsevier, Massachusetts, pp 513–521

Alves RRN, Souto WMS, Barboza RRD (2010) Primates in traditional folk medicine: a world overview. Mammal Rev 40(2):155–180

Andersson TE (2004) Hormones, horses and the menopause industry: the truth about Premarin. Conference paper: American Sociological Association, pp 1–25

Association of British Veterinary Acupuncturists (2018) What conditions can it treat? http://www.abva.co.uk/pet-owner-area/what-conditions-can-it-treat/. Accessed 23 Apr 2018

Athiyaman A (2008) An exploration into modeling sustainable consumption: the case of animal-based tradtional medicines. Proc Acad Mark Stud 13(1):11

Bear Conservation (2018) Bear farms. http://www.bearconservation.org.uk/bear-farms/. Accessed 04 May 2018

Bentham J (1823) Introduction to the principles of morals and legislation. http://www.efm.bris.ac.uk/het/bentham/government.htm. Accessed 20 Mar 2018

Bergenstrahle AE, Nielsen BD (2015) 29 Attitude and behavior of veterinarians surrounding the use of complementary and alternative veterinary medicine in the treatment of equine musculoskeletal pain. J Equine Vet Sci 35(5):395

Borah MP, Prasad SB (2017) Ethnozoological study of animals based medicine used by traditional healers and indigenous inhabitants in the adjoining areas of Gibbon Wildlife Sanctuary, Assam, India. J Ethnobiol Ethnomed 13(1):39

British Association of Veterinary Herbalists (2018) Common conditions that could be treated with herbal supplements. http://www.herbalvets.org.uk/2016/common-conditions-that-could-be-treated-with-herbal-supplements/. Accessed 21 Apr 2018

Challem J (2007) "Bear bile" constituent may help patients who have retinal degeneration. Altern Complement Ther 13(2):113

Costa-Neto EM (1999) Healing with animals in Feira de Santana City, Bahia, Brazil. J Ethnopharmacol 65(3):225–230

Council for International Organization of Medical Sciences and the International Council for Laboratory Animal Science (2012) International guiding principles for biomedical research involving animals. http://iclas.org/wpcontent/uploads/2013/03/CIOMS-ICLAS-Principles-Final.pdf. Accessed 13 Oct 2018

Darwin C (1838) Notebook C, Transmutation of species. http://darwin-online.org.uk/EditorialIntroductions/vanWyhe_notebooks.html. Accessed 26 Jun 2017

Darwin C (1871) The descent of man. http://darwin-online.org.uk/content/frameset?pageseq=1&itemID=F937.1&viewtype=text. Accessed 13 Jan 2018

Das D (2015) Ethnozoological practices among tribal inhabitants in Khowai district of Tripura, North-East India. J Global Biosci 4(9):3364–3372

Department of Health (2013) Medicines/pharmaceuticals of animal origin. Australian Department of Health, Queensland

Electronic Medicines Compendium (2018) Ursodeoxycholic acid. https://www.medicines.org.uk/emc/product/7253/smpc. Accessed 07 May 2018

European Commission (2018) Animals in scientific research. http://ec.europa.eu/environment/chemicals/lab_animals/index_en.htm. Accessed 03 May 2018

Euorpean Union (2012) Commission regulation (EU) No 722/2012. vol L 212

Ferreira F, Fernandes-Ferreira H, Léo Neto N, Brito S, Alves R (2013) The trade of medicinal animals in Brazil: current status and perspectives. Biodivers Conserv 22(4):839–870

Ferreira FS, Brito SV, de Oliveira Almeida W, Alves RRN (2015) Conservation of animals traded for medicinal purposes in Brazil: can products derived from plants or domestic animals replace products of wild animals? https://doi.org/10.1007/s10113-015-0767-4

Guillén J, Vergara P (2018) Global guiding principles: a tool for harmonization. In: Laboratory animals, 2nd edn. Elsevier, Massachusetts, pp 1–13

International Union for Conservation of Nature (2011) IUCN red list if threatened species. http://www.iucnredlist.org/ Accessed 14 Feb 2018

Kikuchi R (2012) Captive bears in human-animal welfare conflict: a case study of bile extraction on Asia's bear farms. J Agric Environ Ethics 25(1):55–77

Kim H, Song M-J (2013) Ethnozoological study of medicinal animals on Jeju Island, Korea. J Ethnopharmacol 146(1):75–82

Luo J, Yan D, Zhang D, Feng X, Yan Y, Xiao X, Dong X (2011) Substitutes for endangered medicinal animal horns and shells exposed by antithrombotic and anticoagulation effects. J Ethnopharmacol 136(1):210–216

Mepham B (1995) An ethical matrix for animal production. New Farmer Grow 46:14

Nascimento AL, Lozano A, Melo JG, Alves RR, Albuquerque UP (2016) Functional aspects of the use of plants and animals in local medical systems and their implications for resilience. J Ethnopharmacol 194:348–357

North American Equine Ranching Information Council (2018) About the equine ranching industry. http://www.naeric.org/about.asp?strNav=4. Accessed 22 Mar 2018

Omasa T, Onitsuka M, Kim W-D (2010) Cell engineering and cultivation of Chinese hamster ovary (CHO) cells. Curr Pharm Biotechnol 11(3):233–240

Pavão MSG, Mourão PAS (2012) Challenges for heparin production: artificial synthesis or alternative natural sources? Glycobiology Insights 3:1–6

People for the Ethical Treatment of Animals (2018) Premarin: a prescription for cruelty. http://www.peta.org/issues/animals-used-for-experimentation/animals-used-experimentation-factsheets/premarin-prescription-cruelty/. Accessed 22 Mar 2018

Rangarajan M (2005) India's wildlife history. Orient Blackswan, Telangana

Rastogi S, Kaphle K (2011) Sustainable traditional medicine: taking the inspirations from ancient veterinary science. Evid Based Complement Altern Med (eCAM) 8(1):1–6

Regan T (1987) The case for animal rights. Springer, New York

Russell WMS, Burch RL, Hume CW (1959) The principles of humane experimental technique. Methuen, London

Sajem Betlu AL (2013) Indigenous knowledge of zootherapeutic use among the Biate tribe of Dima Hasao District, Assam, Northeastern India. J Ethnobiol Ethnomed 9:56

Save the Rhino (2018) Poaching for rhino horn. https://www.savetherhino.org/rhino_info/threats_to_rhino/poaching_for_rhino_horn. Accessed 15 Mar 2018

Silver F (2012) Biomaterials, medical devices and tissue engineering: an integrated approach. Springer Science & Business Media, Berlin

Singer P (1975) Animal liberation: a new ethics for our treatment of animals. New York Review, New York

Singer P (1993) Equality for animals? In practical ethics. Cambridge University Press, Cambridge, pp 55–83

Still J (2003) Use of animal products in traditional Chinese medicine: environmental impact and health hazards. Complement Ther Med 11(2):118–122

Williams VL, Whiting MJ (2016) A picture of health? Animal use and the Faraday traditional medicine market, South Africa. J Ethnopharmacol 179:265–273

World Wildlife Fund (2018) Rhino facts. https://www.worldwildlife.org/species/rhino. Accessed 15 Mar 2018

Yeshi K, Morisco P, Wangchuk P (2017) Animal-derived natural products of Sowa Rigpa medicine: their pharmacopoeial description, current utilization and zoological identification. J Ethnopharmacol 207:192–202

Yibin F, Kayu S, Ning W, Kwan-Ming N, Sai-Wah T, Tadashi N, Yao T (2009) Bear bile: dilemma of traditional medicinal use and animal protection. J Ethnobiol Ethnomed 5:1–9

Chapter 6
Ethical Challenges for the Environment from Traditional and Complementary Medicine

Abstract A healthy environment is essential for human (and animal) health. The delivery of healthcare services can have highly detrimental environmental impacts. For instance, the production, transport and storage of medicines may contribute to the depletion of natural resources and, at the same time, accelerate pollution of the environment with by-products and waste materials. Thus, health care that is intended to keep humans and animals healthy also contributes to ill-health. Given the burgeoning need to make health care more sustainable, the impact of T&CM, specifically herbal medicine, upon the environment is examined. Recommendations are given on how harm to the environment can be avoided and/or minimised. Since the wellbeing of the environment is intricately linked to the health of living organisms, an examination of the relationship between the environment and health is included.

Keywords Traditional and complementary medicine · Herbal medicine Environment · Sustainability · Climate change · Degradation · Peak oil Ecosystems · Health

6.1 Humans and the Environment

Human beings form an integral part of the global ecosystem[1] and their actions have led to profound changes in environments around the world (Wong and Candolin 2015). In the last fifty years, human activities have altered the world's ecosystems more than during any other time span in history and the consequences of these actions might well prove to be devastating and irreversible (Rodríguez et al. 2011). Environmental change is not a new phenomenon, and significant periods of environmental change have taken place long before humans became a dominant species. It is broadly agreed that the earth has experienced at least five periods of mass species extinction prior to the evolution of humankind, with the last of these, the Cretaceous-Tertiary extinction, taking place around sixty-five million years ago (Twitchett 2006). Periods

---

[1]Ecosystems are collections of organisms that occur together in space and time and interact with each other and their physical environment (Wong and Candolin 2015).

© The Author(s), under exclusive license to Springer Nature Switzerland AG 2018
K. Chatfield, *Traditional and Complementary Medicines: Are they Ethical for Humans, Animals and the Environment?* SpringerBriefs in Philosophy
https://doi.org/10.1007/978-3-030-05300-0_6

of mass extinction are characterised by extinction rates that are markedly higher than the more normal 'background' extinction rate. With the rapid loss of species we are currently witnessing, estimated to be between 1000 and 10,000 times higher than the background extinction rate (De Vos et al. 2015), there are widespread claims that we are in the midst of a sixth extinction crisis (Ceballos et al. 2015).

The loss of species poses a threat to those that remain, since a reduction in the diversity of species can affect the resilience of the ecosystem. A dynamically variable range of diverse species helps to buffer ecosystems against perturbations and to maintain longer term stability (Wong and Candolin 2015). All human health ultimately depends upon ecosystem services that are made possible by biodiversity (Patz et al. 2012) and the loss of biodiversity can only have a negative effect on efforts to improve health. The WHO have long been unequivocal in their message about the interrelatedness of the environment (World Health Organization 1986) but the interactions between the environment and health are difficult to assess as we do not have a clear understanding of all the relevant causal relationships (Patz et al. 2012).

Undoubtedly, human activities are having a profound impact on the natural world; increasingly unsustainable practices are placing pressure on natural resources to meet the demands of our economies and the needs of a rapidly growing global population (Patz et al. 2012). The relentless pursuit of economic growth and the associated consumerist propensities are held responsible for considerable damage to the environment:

> For all the wonders of modern civilization—with longevity, opportunities and high-tech solutions made available to the comfortably privileged within the contemporary world order—the impact of advanced globalized consumer culture on the environment and on the majority of humanity has been truly staggering (Poland and Dooris 2010).

The assessment of environmental impacts is extremely complex; impacts are multifaceted and many demand specialist analysis. Debate is ongoing about how best to assess environmental consequences in both the short and long-term. Furthermore, the figures and measurements that would be necessary for accurate calculations are simply not available for many activities. However, there are ways in which we can make reasoned estimations and these can be helpful for making comparisons between different activities as well as for the identification of areas in need of improvement.

For my analysis I will be considering three major, human-induced, threats to the planet as described by Poland et al. (2011): ecological degradation, climate change and peak oil. Each of these threats is relevant to the current and future provision of healthcare, for both conventional medicine and for T&CM. Following a brief description of each, and the impacts of conventional care, I will examine how consideration of these factors can reveal important information for the environmental impact of T&CM through the example of herbal medicine.

6.1.1 Ecological Degradation

'Ecological degradation' or 'environmental degradation' are terms used to describe the deterioration of the environment resulting from a range of factors like the depletion of natural resources (such as water, forests, fuels and soil), the direct destruction of environments (such as through the construction of urban areas or farmland, or as a result of modern transport methods) and the eradication of wildlife. For instance, the expansion of agricultural land is considered to be a primary cause of extensive degradation, as natural habitats are directly removed through the development of farms, and the associated effects of irrigation, soil erosion, pesticides and insecticides (Sharma and Sharma 2014). If degradation becomes sufficiently severe, a threshold is crossed beyond which ecosystems are not able to recover and return to their original state (Gao et al. 2011).

The provision of conventional medical interventions and services contributes to ecological degradation in wide-ranging ways, with the most obvious impacts arising from the pharmaceutical industry and its practices. The development and testing of new pharmaceutical drugs can take many years and require significant resources; a large percentage of products will not even reach the large scale production stage (Dickov and Kuzman 2011). For those drugs that are released onto the market, the production normally comprises a series of steps, often involving multiple companies, at a number of production sites. The process begins with the collection and transportation of raw materials, commonly oils, plants or minerals that are required for synthesis of the drugs, other steps in the manufacturing process and packaging materials. In general, the production facilities for these materials are located at different sites from where the drug is manufactured, and there is a risk at each stage of the production process that chemical waste will enter the environment (Larsson 2008).

In addition to direct contamination of the environment by waste arising from the production processes of pharmaceuticals, a growing body of research evidence is highlighting the problems that arise from the discharge of drug contaminated urine and faeces from both humans and animals.[2] Hundreds of drugs can now be detected at significant concentrations in water, sediment and biological samples taken from the environment. Such compounds, which retain their biological activities, include antibiotics, antidepressants, analgesics and cancer chemotherapy compounds (Dietrich 2008). For example, researchers in many countries have reported that ethinyl oestradiol, used in birth control pills, is responsible for the feminization of male fish in rivers (Depledge 2011).

Conventional healthcare practices contribute to extensive ecological degradation. For example, in the UK, the Government's chief medical adviser, Dame Sally Davies, describes the NHS a 'significant polluter'. In her annual report it is revealed the health

---

[2]Pharmaceuticals are also used in very large quantities as veterinary medicines, especially in the agricultural industry.

service creates 590,000 tonnes of waste a year (Davies 2018), much of which ends up in landfill sites that have the potential to pollute the environment and further ecological degradation (Smith 2013).

6.1.2 Climate Change

Over the past 150 years, human activities in industrialised nations have resulted in the release of large amounts of carbon dioxide and other gases into the atmosphere. Extra carbon in the atmosphere is causing global temperatures to rise and provoking subsequent changes in weather and climate. Changes in levels of rainfall have resulted in more floods, droughts, or intense rain in some regions, as well as more frequent and severe heat waves. The earth's oceans and glaciers have also experienced significant changes, and oceans are warming and becoming more acidic, while ice caps are melting and sea levels are rising (United States Environmental Protection Agency 2018).

While there is disagreement between experts about certain specific aspects of climate change, such as the precise thresholds beyond which change becomes irreversible, the precise timing of the major impacts and how quickly we need to implement changes in order to avoid worse case scenarios (Poland et al. 2011), there is a growing consensus regarding the central concerns. As the most recent completed report (AR5) from the Intergovernmental Panel on Climate Change (2014) outline, there is agreement that:

- Each of the last three decades has been successively warmer at the earth's surface than any preceding decade since 1850.
- In the Northern Hemisphere, the period 1983–2012 was the warmest thirty-year period in the last 1400 years.
- Some of the expected changes from climate change will be abrupt, leaving less time for adaption.
- A large proportion of anthropogenic climate change from carbon dioxide emissions is irreversible on a multi-century to millennial time scale.

The potential of an entity[3] to contribute to climate change is assessed through measurement of the harmful gaseous emissions (greenhouse gases) that are released into the atmosphere through activities undertaken by that entity. This has been termed a 'carbon footprint' and, in the UK, carbon footprint calculations are in high demand (Wiedmann and Minx 2008). However, there is currently no consensus as to how to measure or quantify a carbon footprint: some methods measure emissions of carbon dioxide alone while others include other harmful gasses such as methane (Pandey et al. 2011).

The provision of conventional medicine relies upon many different systems, products and services that contribute towards climate change. In the UK, a series of carbon

---

[3] An entity here meaning anything from an individual to a large business or even nation.

footprints for the NHS have been published relating to years 2004, 2007, 2010, 2012 and 2015 aiding the analysis of trends over time, the latest estimate putting the carbon footprint of the NHS in England (2015) at 22.8 million tonnes of carbon dioxide equivalents (MtCO2e) (Sustainable Development Unit 2016). Between 2007 and 2015 the carbon footprint was successfully reduced by 11% but the rate of reduction must be greatly increased if emissions are to meet the Climate Change Act (2008) target of an 80% reduction by 2050.

Some forms of care have particularly high environmental costs. For example, one year of kidney dialysis has been equated to the environmental impacts of seven return flights between London and New York (Naylor and Appleby 2012), and anaesthetic gases have an estimated 130–2000 times the impact on global warming compared with the same weight of carbon dioxide gas (Ishizawa 2011).

6.1.3 Peak Oil

'Peak oil' is a term used to describe the pattern of production of crude oil as one that grows, reaches a maximum (peak), and then gradually declines to zero. Marion King Hubbert, a geoscientist and employee of the Shell Company, first introduced the notion in 1956 (Aleklett 2012). He proposed that the production pattern of oil forms a symmetrical, bell shaped curve with the peak occurring when about half of a non-renewable resource is extracted (Bardi 2009). Hubbert's work sparked debate on the topic of peak oil that continues to this day. While there is agreement that petroleum production will peak, there is disagreement about specifics such as the peak's timing, the shape of the production curve around the peak, and the post peak rate of decline. Estimates regarding the timing of the peak vary but some argue that the peak of world petroleum production is imminent (Schwartz and Parker 2011). In spite of this potentially imminent challenge, the concept of peak oil has to date had virtually no influence on public health or health care delivery policy.

Across the globe, economic growth is largely dependent upon fossil fuels, most importantly oil. As the supply of cheap, crude oil is depleted, oil supplies will become increasingly more expensive and extraction techniques potentially more damaging to the environment. Increasing scarcity and rising costs of petroleum will have far-reaching impacts on health because petroleum is currently fundamental to the provision of healthcare services (Frumkin et al. 2009). Conventional healthcare practice relies upon the ready availability of petroleum in many ways. A large number of pharmaceutical drugs are derived from petrochemicals as are plastics, resins, solvents, textile fibers, lubricants, and cleaners. Much of modern antiseptic practice depends on the use of disposable plastic materials (ibid.).

Aside from the reliance upon petroleum for medical supplies and equipment, a shortage will greatly affect transportation, as petroleum currently accounts for more than 90% of transportation fuel, in the form of petrol, diesel and jet fuel (Energy Information Administration 2017). Shortages in petroleum supply will impact upon functions such as the transport of supplies, the transport of healthcare workers and patients, the provision of ambulances (including air ambulances) and public health services, such as community based health visitors and public health inspectors (Frumkin et al. 2011). It is impossible to predict precisely the extent of the problem, because the impacts are potentially so far-reaching.

Consideration of the effects that the provision of healthcare has on the three major, human-induced, threats to the planet of ecological degradation, climate change and peak oil, serves to exemplify the tension that exists between the wellbeing of humans and the wellbeing of the environment. However, it is also clear that the health and wellbeing of humans is dependent upon a healthy environment and it is in the interests of humans to protect the environment. Paradoxically, healthcare systems that are intended to improve human health also contribute to environmental damage.

The need for all forms of healthcare to mitigate harmful effects caused by their interventions will inevitably increase as natural resources decline and the problems associated with climate change increase. I believe that this will ultimately demand scrutiny of practices down to the finest detail. Given that this is not currently possible, in the following I use information that is available to provide an overview of the main environmental impacts of herbal medicine. This analysis is undertaken from a Western perspective. In other parts of the world, where herbs are used as locally based, indigenous medicine, the associated environmental effects will obviously be quite different.

6.2 Analysing the Environmental Impacts of Western Herbal Medicine

The environmental impacts of T&CM that were revealed in the construction of the ethical matrix related exclusively to the practice of herbal medicine, suggesting that these impacts are greater and/or more obvious than for other forms of T&CM.

Herbal treatments are derived from plants and the wide scale use of plants in medicine has a direct impact upon the environment, as some plant species are in great demand. Most prescribed herbal medicines are administered in the form of a liquid tincture, but patients may also be given herbal tea, tablets, ointments, creams or lotions (Casey et al. 2007). Ethanol is commonly used for the extraction and preservation of herbal products to form tinctures and the tinctures are typically stored in brown glass bottles. Over-the-counter herbal products are most often sold in the form of tablets or capsules in plastic containers (De Bolle et al. 2008). Aside from the production, delivery and storage costs of the herbal medications, environmental impacts also arise from energy consumption for travel (practitioners and patients) and from maintenance of clinics. Hence, the primary environmental impacts of herbal medicine stem from the use of plants, ethanol, glass and plastic, as well as general running of clinic premises. The damaging impacts from each of these is considered, along with suggestions for reduction and/or mitigation, in relation to the three major human threats to the planet: ecological degradation, global warming and peak oil.

6.2.1 Herbal Medicines and Ecological Degradation

Use of Plants

The harvesting of medicinal plants, especially those in high demand, can lead to the degradation of land and the loss of biodiversity (Rastogi and Kaphle 2011). Many plants used for medicinal purposes are under threat of extinction from the destruction and over-collection of plants in their natural habitats (McKenzie et al. 2009). Harvesting without replacement planting, deforestation and the increased marketing of medicinal plants have resulted in the decline and near-extinction of many valuable medicinal plant species around the world (Dahlberg and Trygger 2009). Many medicinal plants are collected from the wild in an uncontrolled manner, and cultivated plants are often considered inferior to their counterparts (Ncube et al. 2012). In the US, the origin of many well known medicinal plants, a significant number have been overcollected almost to the point of extinction in their natural habitats, including the pacific yew, ginseng, goldenseal, black cohosh, and echinacea (McKenzie et al. 2009). The high demand for some products can motivate producers to engage in mass production without regard for the environmental consequences or for the finer details of plant cultivation (van Andel and Havinga 2008).

On the other hand, many plants are awarded value because of their role in T&CM (Timmermans 2003) and direct local use contributes to the preservation of species and habitats (Brown 1995). Wild harvest gives an economic value to ecosystems and habitats and thus provides an incentive for protection. The involvement of local people in sustainable management practices increases both their desire and their ability to protect wild populations from over-exploitation (McKenzie et al. 2009).

Reduction and/or Mitigation

The environmental costs associated with herbal medicine are recognised by many in the herbal medicine field and some action is already being taken. For example, Botanic Gardens Conservational International (2018), a membership organisation representing a network of 500 botanic gardens in more than 100 countries, is working to secure the future of the world's most threatened medicinal plants. They have identified a multi-method approach to conservation, based upon targeted research, education, collaboration, conservation, preservation of indigenous knowledge, developing alternatives to wild harvesting and ensuring sustainable wild harvests (Hawkins 2008).

Use of Ethanol

Ethanol is commonly used for the extraction and preservation of herbal products to form tinctures and for herbal medicine this bioethanol is produced by the fermentation of natural sugars with yeast. Most of the world's bioethanol is produced from crops such as sugar cane, sugar beet, corn, rice and maize. Changes in land use, where crops are grown for ethanol production, can cause substantial ecological degradation through an increase in tillage[4] intensity, use of fertilizers and pesticides, depletion of soil minerals, water-induced soil erosion and greenhouse gas emissions (Larson

---

[4]*Tillage* is the agricultural preparation of soil by mechanical agitation of various types.

et al. 2010). In the UK, bioethanol production only began in 2007, but there are now several industrial sites producing ethanol from sugar beet and wheat (Alberici and Toop 2013). Still, much of the ethanol used in the UK is imported from Europe or the US where the primary source material is corn (Hofstrand 2009a).

Much of the electricity used in the production of ethanol comes from fossil fuels, with only a small portion coming from renewable sources such as wind (Hofstrand 2009b).

Reduction and/or Mitigation
Advances in biotechnology permit the production of ethanol from waste biomass using bacteria. Engineered strains of *Escherichia coli* can convert plant sugars from waste into ethanol while the main crop can be grown as a food source. The use of waste to produce ethanol may also help with disposal and reduce the practice of burning crop residues (Larson et al. 2010).

Use of Glass
There are many different types of glass, but the type most commonly used for the production of bottles and jars is composed of three main components; silica sand, soda ash and limestone. The use of glass bottles leads to a number of detrimental environmental impacts from transportation of the raw materials, production of the glass containers, transportation to the pharmacy, transportation of the filled bottles to the consumer, transportation of the empty bottles for disposal or recycling. Glass manufacturing is among the most energy-intensive of industries (United States Energy Information 2013). The production stage of the glass bottle has the highest environmental impact because very high temperature furnaces (up to 1200 degrees centigrade) are used and have very high energy requirements for the melting of raw materials to form glass. Road transport has the second highest effect and is higher than for other types of containers or packaging because of the relative weight of the glass (Banar and Cokaygil 2008). In addition, there can be localised environmental degradation around glassworks arising from fresh water use, water pollution and dust (Gander 2008).

Reduction and/or Mitigation
On the positive side, glass is totally recyclable and can be recycled an infinite number of times without loss of quality (Blengini et al. 2012). The use of recycling avoids landfill disposal and, for a glass manufacturer, the use of cullet (recycled glass) is extremely beneficial. Aside from savings in virgin raw material consumption, around 3% in energy savings can be achieved for every 10% of cullet that replaces these raw materials, because no 'reaction energy' is needed to melt cullet (British Glass 2018).

Use of Plastic
Most plastics are petroleum based. They are cost-effective, versatile, lightweight and an ideal material for many disposable applications. In 2015, global plastic production reached 322 million tonnes (Narancic and O'Connor 2017) and about 50% of global production is used for disposable applications, products that are discarded within a year of their purchase (Halden 2010). The disposal of plastics is a major cause

of ecological degradation. The utility of plastic in contemporary society is undeniable but the perceived benefit of single use, throw away products and packaging is accompanied by persistent waste as plastics degrade extremely slowly. Disposal of plastics in landfills diminishes land resources (North and Halden 2013) and a significant amount of disposable items, such as plastic bags, bypass legitimate disposal and enter the environment, resulting in widespread, long-term pollution. This is not just a threat to the land: plastic pollution is now ubiquitous in aquatic environments, posing a major threat to aquatic life (Eriksen 2014).

Reduction and/or Mitigation
In theory, the recycling of plastics represents a plausible solution to many of these problems but, in practice, there are numerous logistical challenges, including the lack of effective sorting techniques, since the mixing of different plastic source materials has a significant impact upon the quality of the resultant recycled materials. Increasing the use of biodegradable plastics could also help to reduce impacts upon ecological degradation; however, it can do so sustainably only if these alternatives are made from non-fossil resources using renewable energy (North and Halden 2013).

6.2.2 Herbal Medicines and Climate Change

Use of Plants
The growing and harvesting, processing and packaging, and transport of herbs are all energy intensive; the carbon footprint associated with production and transport contributes to climate change that in turn may lead to the extinction of species, reduction in their availability and also reduction in the quality of the resultant products (Rastogi and Kaphle 2011).

Many herbs are harvested from diverse geographic locations and the transport requirements from source to market may be significant, including transport from source to raw material wholesalers, to manufacturers for processing, to wholesale warehouses and finally to practitioners, clinics and retail outlets (McElroy 2011). Similarly, the processing of herbs may have many different requirements depending upon the final product, including drying, milling, encapsulating, pressing and percolation (*ibid.*), each of which has requires energy that is commonly derived from fossil fuels.

Reduction and/or Mitigation
The use of herbal medications that can be grown and processed locally should be prioritised.

Use of Ethanol
The production of bioethanol is a source of greenhouse gasses at various different points along the life cycle. Carbon dioxide emissions result from the combustion of fossil fuels on the farm that are directly related to crop production. Off the farm, carbon dioxide emissions are released through the use of fossil fuels for the manufacturing and transportation of inputs such as fertilizers, pesticides and seeds (Larson

et al. 2010). Crop production also results in a substantial amount of nitrous oxide, a powerful greenhouse gas, estimated to be 289 times as powerful as carbon dioxide. The large increase in the use of nitrogen fertilizer for producing high nitrogen consuming crops like corn has increased nitrous oxide emissions (Hofstrand 2009a). In addition, the manufacture of fertilizer and lime is also a large source of nitrous oxide emissions (Hofstrand 2009b).

Reduction and/or Mitigation
A switch to renewable sources of energy and a reduction in the reliance upon nitrogen fertilizers are needed to reduce the impact that ethanol production has upon climate change.

Use of Glass
The major contributors to climate change arising from the production of glass are the carbon dioxide emissions that result from the combustion of the fuels used to fire the furnaces and from associated road transport (Nilsson et al. 2011). In addition, carbon dioxide is also produced by the decomposition of raw materials during the melting. This is the only greenhouse gas emitted during the production of glass (AGC Glass Europe 2018).

Reduction and/or Mitigation
Recycling can reduce these carbon dioxide emissions, just as the use of cullet decreases the energy required in production and the amount of carbon dioxide released during melting (British Glass 2018).

Use of Plastic
The production of new plastics is energy intensive. It has been estimated that the production of 1 kg of plastics from crude oil requires between 62 and 108 megajoules of energy, compared with 20–25 megajoules for iron (from iron ore); 18–35 megajoules for glass (from sand) and 20–50 megajoules for steel (De Decker 2014). The resulting carbon footprint for plastic is around six kilograms of carbon dioxide per kilogram produced (Rohrer 2011). Additionally, disposal through incineration results in the release of high levels of carbon dioxide and other greenhouse gases (North and Halden 2013).

Reduction and/or Mitigation
Plastic pollution in the environment is currently receiving worldwide attention (Chae and An 2018) but suggestions for alternatives must be considered for their impacts upon climate changes as well as their contribution to degradation.

Maintenance of Clinics
Aside from the equipment used, the running of premises and the travel needs of practitioners and patients all contribute to climate change through the gaseous emissions arising from energy production and the use of fuel. As with all forms of T&CM in the UK, practitioners work predominantly in private practice, often working alone. The relative number of medical herbalists to the general population is very small when compared with the number of General Practitioners in the UK and people may have to travel some distance for treatment, because practitioners are not evenly distributed across the country.

Reduction and/or Mitigation
Central location of practitioners within community health clinics and increased use of telemedicine are just two possible solutions.

6.2.3   Herbal Medicines and Peak Oil

The transportation needed for the production and delivery of all materials used in herbal medicine, as well as the travel of patients and practitioners, all rely upon crude oil based fuels. In particular, the packaging of herbs is oil intensive. Additionally, if oil prices rise, this will lead to the decreased availability of herbs from overseas and increased costs of transport and packaging (McElroy 2011).

Of particular concern is that the current production of plastics from crude oil is not sustainable. Given the declining reserves of oil (and other fossil fuels), its use for the manufacture of non-recyclable goods is not sustainable. There is pressing need to increase the recycling capacity of products and develop suitable plastics (or plastic alternatives) from renewable sources (Sabaliauskaitė and Kliaugaitė 2014).

Reduction and/or Mitigation

Strategies that are adopted for reduction and/or mitigation of degradation and climate change also, importantly, will reduce dependence upon oil.

6.2.4   In Summary

The environmental impacts of herbal medicine may be damaging in some respects but it appears that measures can be taken to reduce or mitigate these successfully. Given that a large percentage of the global population rely upon herbal medicines for their primary healthcare it is essential that this is addressed for the health and wellbeing of humans and the environment. Some herbal medicine producers are already incorporating these mitigating measures. For example, Rutland Biodynamics, a UK producer of organic herbal products, aims for complete sustainability. To this end they conduct a total audit of the environmental impact of all activities, from raw materials (land, seeds and packaging) through to composting and drainage (Rutland Biodynamics 2015). They take proactive measures to offset their contributions to environmental damage, such as ensuring that everything they send out is recyclable, and the planting of thousands of new trees. It is now estimated that their activities absorb more greenhouse gases than are generated (*ibid.*).

6.3   Future-Proofing T&CM

In the light of immense challenges facing the current and future provision of health care, and the urgent need to reduce the environmental fallout, the issue of whether T&CM can contribute to a more environmentally friendly healthcare system is worthy of consideration. According to the WHO, the predominance of curative, hospital-based, disease-oriented services are 'top-heavy' and responsible for huge inefficiencies that could be redirected towards achieving universal health coverage (World Health Organization 2013). It is broadly agreed that prevention must be at the core of sustainable health care; the most sustainable approach being one that minimises care needs by preventing ill-health and supporting people to manage their own health as effectively as they can (Naylor and Appleby 2013). Here, there is obvious congruence with the underlying philosophies of many forms of T&CM, as well as with the notion that health is rooted in the ability to adapt and respond.

The current provision of conventional medicine is vastly more damaging than the provision of T&CM in terms of energy use, reliance upon oil and pollution of the environment, but direct comparison is not feasible for all services. For example, there is no form of T&CM that could serve as an alternative to the intensive care treatment required for those who have suffered traumatic injuries. However, there are some potentially comparable areas. For instance, the environmental impacts of osteopathic and chiropractic approaches to the treatment of musculoskeletal problems can be directly compared with a conventional approach. Chiropractors and osteopaths rely upon few resources, other than their own skills, for the treatment of patients. Their primary piece of equipment is a treatment table and these have a very long life span. Additionally, use of recycled steel and natural fibres in the tables, helps to reduce environmental damage even further. Osteopathy and chiropractic treatments could potentially reduce the need for pharmaceutical drugs, hospital stays and surgical interventions (Erwin et al. 2013) thereby reducing both economic and environmental costs. Homeopathic medications, treatment of choice for many all over the world, make use of ethanol and glass but in much smaller quantities than herbal medications. Additionally, because of the way in which homeopathic medicines are made and prescribed, only tiny amounts of source materials are required for the supply of hundreds or even thousands of practitioners (Morgan 2014). The production and disposal of acupuncture needles is a more challenging concern. The current requirement for sterile and single use needles has implications for the consumption of natural resources (steel, copper and oil), the production of greenhouse gases and pollution of the environment arising from disposal. The lack of recycling of used acupuncture needles results in a permanent loss of these materials as a resource; this practice is a problem for the future sustainability of acupuncture and requires attention.

Of course, decisions about health care options cannot be based purely upon their environmental impacts; equally, they should not be based purely upon efficacy and/or economic costs without consideration of their environmental burden. Ultimately, for humankind, there should be no conflict between the wellbeing of humans and the wellbeing of the environment. Respect for the wellbeing of the environment is fundamental for human health and health care services that damage the environment, also serve to damage the potential for human health and wellbeing.

References

AGC Glass Europe (2018) Our environmental impact. http://www.agc-glass.eu/en/sustainability/environmental-achievements/environmental-impact. Accessed 21 Nov 2017

Alberici S, Toop G (2013) UK Biofuels industry overview. Department of Transport, London

Aleklett K (2012) Peak oil. In: Peeking at peak oil. Springer, New York, pp 7–15

Banar M, Cokaygil Z (2008) A comparative life cycle analysis of two different juice packages. Environ Eng Sci 25(4):549–556

Bardi U (2009) Peak oil: the four stages of a new idea. Energy 34(3):323–326

Blengini GA, Busto M, Fantoni M, Fino D (2012) Eco-efficient waste glass recycling: integrated waste management and green product development through LCA. Waste Manage 32(5):1000–1008

Botanic Gardens Conservation International (2018) Mission and strategy. https://www.bgci.org/about-us/mission/. Accessed 02 May 2018

British Glass (2018) Rycycling. https://www.britglass.org.uk/our-work/recycling Accessed 17 Jan 2018

Brown K (1995) Medicinal plants, indigenous medicine and conservation of biodiversity in Ghana. Cambridge University Press, Cambridge

Casey M, Adams J, Sibbritt D (2007) An examination of the prescription and dispensing of medicines by Western herbal therapists: a national survey in Australia. Complement Ther Med 15(1):13–20

Ceballos G, Ehrlich PR, Barnosky AD, García A, Pringle RM, Palmer TM (2015) Accelerated modern human-induced species losses: entering the sixth mass extinction. Sci Adv 1(5):e1400253

Chae Y, An YJ (2018) Current research trends on plastic pollution and ecological impacts on the soil ecosystem: a review. Environ Pollut 240:387–395

Climate Change Act (2008) UK. https://www.legislation.gov.uk/ukpga/2008/27/contents. Accessed 12 Oct 2018

Dahlberg AC, Trygger SB (2009) Indigenous medicine and primary health care: the importance of lay knowledge and use of medicinal plants in rural South Africa. Hum Ecol: An Interdisc J 37(1):79–94

Davies S (2018) Annual report of the Chief Medical Officer 2017: health impacts of all pollution. Department of Health and Social Care, London

De Bolle L, Mehuys E, Adriaens E, Remon J-P, Van Bortel L, Christiaens T (2008) Home medication cabinets and self-medication: a source of potential health threats? Ann Pharmacother 42(4):572–579

De Decker K (2014) How much energy does it take? http://www.lowtechmagazine.com/what-is-the-embodied-energy-of-materials.html. Accessed 06 Sept 2017

Depledge M (2011) Does the pharmaceutical industry need a new prescription? Sci Parliam 68(4):44

De Vos JM, Joppa LN, Gittleman JL, Stephens PR, Pimm SL (2015) Estimating the normal background rate of species extinction. Conserv Biol 29(2):452–462

Dickov V, Kuzman B (2011) Analyzing pharmaceutical industry. Natl J Physiol Pharm Pharmacol 1(1):1–8

Dietrich AS (2008) Corrosion in the system: the community health by-products of pharmaceutical production in Northern Puerto Rico. In: Singer M, Baer H (eds) Killer commodities: public health and the corporate production of harm. Altimera Press, Plymouth, pp 335–365

Energy Information Administration (2017) Energy use for transportation. http://www.eia.gov/Energyexplained/?page=us_energy_transportation. Accessed 08 July 2017

Eriksen M (2014) The plastisphere—the making of a plasticized world. Tulane Environ Law J 27(2):153–163

Erwin WM, Korpela AP, Jones RC (2013) Chiropractors as primary spine care providers: precedents and essential measures. J Can Chiropr Assoc 57(4):285–291

Frumkin H, Hess J, Parker CL, Schwartz BS (2011) Peak petroleum: fuel for public health debate. Am J Public Health 101(9):1542

Frumkin H, Hess J, Vindigni S (2009) Energy and public health: the challenge of peak petroleum. Public Health Rep 124(1):5–19

Gander P (2008) Glass half full for availability, price and carbon footprint. Food Manuf 83(11):23

Gao Y, Zhong B, Yue H, Wu B, Cao S (2011) A degradation threshold for irreversible loss of soil productivity: a long-term case study in China. J Appl Ecol 48(5):1145–1154

Halden RU (2010) Plastics and health risks. Annu Rev Public Health 31:179–194

Hawkins B (2008) Plants for life: medicinal plant conservation and botanic gardens. Botanic Gardens Conservation International, Surrey

Hofstrand D (2009a) Efficiency and environmental improvements of corn ethanol production. http://www.agmrc.org/renewable_energy/ethanol/efficiency-and-environmental-improvements-of-corn-ethanol-production. Access 12 Mar 2018

Hofstrand D (2009b) Greenhouse gas emissions of corn ethanol production. http://www.agmrc.org/renewable_energy/climate_change/greenhouse-gas-emissions-of-corn-ethanol-production/. Accessed 12 Mar 2018

Intergovernmental Panel on Climate Change (2014) Climate change 2014: impacts, adaption and vulnerability. http://www.ipcc.ch/report/ar5/wg2/. Accessed 30 Mar 2018

Ishizawa Y (2011) General anesthetic gases and the global environment. Anesth Analg 112(1):213–217

Larson J, English B, Ugarte DDLT, Menard R, Hellwinckel C, West TO (2010) Economic and environmental impacts of the corn grain ethanol industry on the United States agricultural sector. J Soil Water Conserv 65(5):267–279

Larsson D (2008) Drug production facilities—an overlooked discharge source for pharmaceuticals to the environment. In: Pharmaceuticals in the environment. Springer, Berlin, pp 37–42

McElroy K (2011) Herbal medicine practice: future environmental impacts. Aust J Med Herbalism 23(4):164–167

McKenzie M, Kirakosyan A, Kaufman PB (2009) Risks associated with over collection of medicinal plants in natural habitats. In: Recent advances in plant biotechnology. Springer, Berlin, pp 363–387

Morgan J (2014) The carbon footprint of homeopathy [personal communication 17/11/2014]

Narancic T, O'Connor KE (2017) Microbial biotechnology addressing the plastic waste disaster. Microb Biotechnol 10(5):1232–1235

Naylor C, Appleby J (2012) Sustainable health and social care: connecting environmental and financial performance. King's Fund, London

Naylor C, Appleby J (2013) Environmentally sustainable health and social care: scoping review and implications for the English NHS. J Health Serv Res Policy 18(2):114–121

Ncube B, Finnie JF, Van Staden J (2012) Quality from the field: the impact of environmental factors as quality determinants in medicinal plants. S Afr J Bot 82:11–20

Nilsson K, Sund V, Florén B (2011) The environmental impact of the consumption of sweets, crisps, and soft drinks. Nordic Council of Ministers, Copenhagen

North EJ, Halden RU (2013) Plastics and environmental health: the road ahead. Rev Environ Health 28(1):1–8

Pandey D, Agrawal M, Pandey JS (2011) Carbon footprint: current methods of estimation. Environ Monit Assess 178(1–4):135–160

Patz J, Corvalan C, Hortwitz P, Campbell-Lendrum D (2012) Our planet, our health, our future. Human health and the Rio Conventions: biological diversity, climate change and desertification. World Health Organization, Geneva

Poland B, Dooris M (2010) A green and healthy future: the settings approach to building health, equity and sustainability. Critl Public Health 20(3):281–298

Poland B, Dooris M, Haluza-Delay R (2011) Securing 'supportive environments' for health in the face of ecosystem collapse: meeting the triple threat with a sociology of creative transformation. Health Promot Int 26(2):ii202–ii215

Rastogi S, Kaphle K (2011) Sustainable traditional medicine: taking the inspirations from ancient veterinary science. Evid Based Complement Altern Med (eCAM) 8(1):1–6

Rodríguez JP, Rodriguez-Clark KM, Baillie JE, Ash N, Benson J, Boucher T, Brown C, Burgess ND, Collen B, Jennings M (2011) Establishing IUCN red list criteria for threatened ecosystems. Conserv Biol 25(1):21–29

Rohrer J (2011) Plastic bags and plastic bottles: CO2 emissions during their lifetime. https://timeforchange.org/plastic-bags-and-plastic-bottles-CO2-emissions. Accessed 12 Oct 2018

Rutland Biodynamics (2015) Environmental impact of herbal medicine. http://www.rutlandbio.com/about-rutlandbio/environmental-impact/. Accessed 30 Apr 2018

Sabaliauskaitė K, Kliaugaitė D (2014) Resource efficiency and carbon footprint minimization in manufacture of plastic products. Efektyvus išteklių naudojimas ir anglies dioksido pėdsako mažinimas plastikinių gaminių gamyboje. 67(1):25–34

Schwartz BS, Parker CL (2011) Public health and medicine in an age of energy scarcity: the case of petroleum. Am J Public Health 101(9):1560

Sharma M, Sharma K (2014) Food security and ecological degradation: challenges and opportunities for inclusive growth. Am J Public Health 3(200):2

Smith K (2013) Environmental hazards: assessing risk and reducing disaster. Routledge, London

Sustainable Development Unit (2016) Carbon footprint update for NHS England. https://www.sduhealth.org.uk/policy-strategy/reporting/nhs-carbon-footprint.aspx. Accessed 20 Feb 2018

Timmermans K (2003) Intellectual property rights and traditional medicine: policy dilemmas at the interface. Soc Sci Med 57(4):745–756

Twitchett RJ (2006) The palaeoclimatology, palaeoecology and palaeoenvironmental analysis of mass extinction events. Palaeogeogr Palaeoclimatol Palaeoecol 232(2):190–213

United States Energy Information (2013) Today in energy. http://www.eia.gov/todayinenergy/detail.cfm?id=12631. Accessed 30 Oct 2017

United States Environmental Protection Agency (2018) Climate indicators. https://www.epa.gov/climate-indicators/oceans. Accessed 03 May 2018

van Andel T, Havinga R (2008) Sustainability aspects of commercial medicinal plant harvesting in Suriname. For Ecol Manage 256(8):1540–1545

Wiedmann T, Minx J (2008) A definition of 'carbon footprint'. Ecol Econ Res Trends 1:1–11

Wong BBM, Candolin U (2015) Behavioral responses to changing environments. Behav Ecol 26(3):665–673

World Health Organization (1986) Ottawa charter for health promotion. World Health Organization, Geneva

World Health Organization (2013) WHO traditional medicine strategy: 2014–2023, Hong Kong SAR: World Health Organization

Chapter 7
Is Traditional and Complementary Medicine Ethical?

Abstract All forms of health care are associated with ethical challenges. This chapter briefly summarises an endeavour to identify, analyse and make recommendations for addressing challenges that are associated with traditional and complementary medicine (T&CM). This necessitated careful consideration from multiple perspectives, reflection upon what 'health' means and comparisons with conventional biomedicine. For many, T&CMs can be regarded as a key component of their 'resilience toolkit'; when health is threatened, use of T&CMs can help people to respond positively.

Keyword Traditional medicine · Complementary medicine · Resilience toolkit Ethical challenges · Healthcare ethics

Which ethical concerns are related to the use of T&CMs? To answer this question, two clarifications were necessary.

First, a distinction had to be made between traditional or indigenous forms of healing that are rooted in local culture and those which are used in a complementary or alternative manner. For instance, acupuncture and the use of Chinese herbal medicines are classified as traditional medicine when used within China, but as complementary or alternative medicine elsewhere in the world.

Secondly, the meaning of 'health' had to be defined to ensure that reasoned judgements about which health-related activities are ethical (or not) could be made. Health is not just about the absence of symptoms and neither is it about complete wellbeing, as is stated in the most widely referenced definition from the WHO (1946): "Health is a state of complete physical, mental and social wellbeing and not merely the absence of disease or infirmity".

Consideration of the healthiest populations in the world painted a vibrant picture of what health can look like, and enabled insight into how it might be achieved. The high levels of health that are experienced by the Okinawans in Japan, for instance, do not stem from access to expensive, high tech medical systems, but rather, their health can be attributed to a variety of simpler measures, such as a healthy diet, staying active into old age and having a strong sense of community. This community spirit, or 'yuimaru', as it is termed in Okinawa, is widely credited as an important factor for longevity and health into old age (Willcox et al. 2013).

Unlike the WHO definition of health, the definition proposed by Huber et al. (2011): "[Health is] the ability to adapt and to self-manage when facing physical, mental, and social challenges" (p. 2) reflects the resilience of the Okinawans who, in spite of their exposure to potentially stressful situations, exhibit low levels of disease, and live long and active lives (Willcox et al. 2008; Buettner 2012). This more contemporary definition of health also reflects a convergence of thinking across various disciplines, such as psychology and environmental science, and is in direct alignment with the way in which many T&CM philosophies describe health. For this reason, it was adopted for the research undertaken in this book.

Over the past 30 years, many ethical challenges for T&CM have been described in the literature. Some authors are particularly vociferous in their condemnation of T&CM practice and practitioners. To assess a question as broad as "Which ethical concerns are related to the use of T&CMs?", a reliable methodology had to be found; a methodology which would help classify and assess individual ethical challenges. The most suitable methodology was the ethical matrix (Mepham 2005). Implementation of the matrix, facilitated categorisation of a wide range of ethical concerns. The need to include a number of stakeholders in the ethical matrix, broadened the analysis to incorporate animals and the environment. It thus allowed the analysis of ethical challenges for each of these stakeholders. The remainder of this chapter summarises the results of this analysis.

7.1   Ethical Challenges for Humans

Development of an ethical matrix for T&CM revealed that most published ethical concerns were related to human users. Identified challenges included the potential for side effects from medicines; injuries from treatments; inadequate informed consent procedures; and harm from refusal or delayed access to conventional biomedical care. Underpinning these challenges are assertions that there is no robust evidence base for T&CM (as is required for new conventional treatments), as well as poor standards in education and regulation of practitioners. Given the vast array of identified ethical challenges related to humans, analysis of ethical concerns for human users was restricted to issues related to safety.

Comparisons of the potential for adverse drug reactions (ADRs) and adverse events (AEs) with conventional medicine were illuminating. In spite of the extensive testing of conventional drugs, and training and regulation of medical professionals, ADRs and AEs happen in conventional medicine on a routine basis, in alarming high numbers, as highlighted back in 2000 with Kohn's report, '*To Err is Human*', regarding preventable deaths in US hospitals (Kohn et al. 2000). This is a problem for all medicine, not just T&CM interventions and steps need to be taken across all domains to help improve safety. Pragmatic steps for improving safety in T&CM in relation to ADRs from herbal medicine and AEs associated with homeopathy were proposed. In particular, the availability of unlicensed, unregulated and potentially lethal medications, via the internet requires global attention and cooperation.

7.2   Ethical Challenges for the Environment

Climate change is broadly considered the greatest challenge to healthcare of the twenty-first century (Costello et al. 2009). Together with increasing ecological degradation and the inevitable decline in natural resources such as oil, these are major threats to the environment and to the health of all living beings. There is an urgent need for health care services to respond to these threats and to prepare for the challenges that the future holds (Naylor and Appleby 2013). Conventional healthcare services contribute to environmental damage on a scale that is unsustainable (Verkerk 2009), and there is a need for all aspects of health and social care to examine their own contributions to such damages in a critical fashion. From an environmental perspective, most T&CMs have a relatively minor damaging impact that can be reduced and/or mitigated to help minimise harm and ensure sustainability of these T&CMs. The sourcing of herbal medicine products poses challenges for loss of diversity and ecological degradation, but steps are being taken on an international scale through organisations such as Botanic Gardens Conservation International to minimise this damage (Hawkins 2008).

7.3   Ethical Challenges for Animals

The use of animals in the production of T&CM products is a cause of great concern. While animals are also used in conventional medicine, use in T&CM is more commonplace, it affects more species, including endangered species, and in many regions, is not subject to any regulation. The use of protected animal species in T&CM products is particularly worrisome but this does not pose a dilemma for ethical analysis because it is simply wrong. However, demand for animal use in T&CM products has been the main contributory factor in the declining numbers of many animals such as the rhino and the tiger and this demand continues, in spite of legal protection. Given that zootherapy[1] is most common in areas where people have poor or limited access to conventional medications, the use of animals in medications is unlikely to be simply a matter of choice. As well as providing the only means of healthcare, zootherapy may also be deeply rooted in local culture and traditions. Nevertheless, exploration of alternatives and of ethical use are warranted and here, application of the principles of replacement, reduction and refinement, may prove to be of benefit, just as they have been for animal use in research.

---

[1]The use of animal-based products for medicinal purposes.

7.4   Conclusion

All forms of healthcare interventions are associated with ethical challenges; they can be related to the practitioners, the intervention itself, the sourcing and production, or the effects upon the environment. Hence, it is unsurprising that the analysis in this book revealed ethical challenges for all stakeholders. T&CM can be practised ethically and it can also be practised unethically. This is no different from conventional medicine.

Conventional medicine tends to be better regulated than T&CM and thus higher ethical standards can be assumed. Undoubtedly, ethical standards in T&CM would benefit from improved and harmonious global governance. This is especially important to avoid overuse of T&CM plant or animal-based products, to avoid exploitation of animals and environments, and to protect humans from unregulated medicines. However, the superior regulation of conventional medicine is no assurance of higher ethical standards in all matters.

Comparisons with conventional medicine reveal that T&CM fares especially well in particular areas. The environmental impacts from T&CM are minor in comparison to those from conventional healthcare. The current provision of conventional medicine is vastly more damaging than the provision of most complementary medicines in terms of energy use, reliance upon oil and pollution of the environment, although direct comparison is not feasible for all services. For example, T&CM cannot serve as an alternative to the intensive care treatment required for those who have suffered traumatic injuries. However, there may be many potentially fruitful areas where conventional medicine is struggling, such as with antibiotic resistance, or multimorbidity in the elderly. Additionally, it is broadly agreed that prevention must be at the core of sustainable healthcare; the most sustainable approach being one that minimises care needs by preventing ill-health and supporting people to manage their own health as effectively as they can (Naylor and Appleby 2013).

In the literature, reference is made increasingly to the notion of a 'resilience toolkit'; one that contains the tools for ensuring resilience. This has been applied in a number of fields, including community resilience[2] (Pfefferbaum et al. 2015), mental health resilience (Southwick 2011) and climate change resilience (Engle et al. 2014). It is premised upon the assumption that, for effective resilience or adaption, people need to have the necessary resources in place.

If health is the ability to adapt and to self-manage when facing physical, mental, and social challenges, it should be possible to imagine what a resilience toolkit for health might look like, to question which resources are the most vital and to examine the place of T&CM within that toolkit. Obviously, the basic building blocks for health; food, clean water, shelter and so on, will be essential ingredients of the toolkit, as well as relevant public health interventions. Aside from this, the Okinawans might tell us we need a good deal of community spirit, to stay active, and to maintain our

---

[2]Community resilience can be thought of as 'local disaster readiness'.

autonomy into old age. For some, their resilience toolkit may include T&CM if it can be used as a resource that helps them to stay well. The challenge for legislators and policy-makers is to ensure that this resource can be used in a safe manner, without undue harm to humans, animals and the environment.

References

Buettner D (2012) The Blue Zones: 9 lessons for living longer from the people who've lived the longest. Geographic Books, Washington

Costello A, Abbas M, Allen A, Ball S, Bell S, Bellamy R, Friel S, Groce N, Johnson A, Kett M (2009) Managing the health effects of climate change: Lancet and University College London Institute for Global Health Commission. The Lancet 373(9676):1693–1733

Engle N, Bremond A, Malone E, Moss R (2014) Towards a resilience indicator framework for making climate-change adaptation decisions. Mitig Adapt Strat Glob Change 19(8):1295–1312

Hawkins B (2008) Plants for life: medicinal plant conservation and botanic gardens. Bot Gard Conserv Int, Surrey

Huber M, Knottnerus JA, Green L, van der Horst H, Jadad AR, Kromhout D, Leonard B, Lorig K, Loureiro MI, van der Meer JWM, Schnabel P, Smith R, van Weel C, Smid H (2011) How should we define health? BMJ 343:d4163

Kohn LT, Corrigan JM, Donaldson MS (2000) To Err Is Human: building a safer health system. National Academies Press, Washington

Mepham B (2005) A framework for ethical analysis. Bioethics: an introduction for the biosciences. Oxford University Press, Oxford

Naylor C, Appleby J (2013) Environmentally sustainable health and social care: scoping review and implications for the English NHS. J Health Serv Res Policy 18(2):114–121

Pfefferbaum RL, Pfefferbaum B, Nitiéma P, Houston JB, Van Horn RL (2015) Assessing community resilience. Am Behav Sci 59(2):181–199

Southwick SM (2011) Resilience and mental health: responding to challenges across the lifespan. Cambridge University Press, Cambridge

Verkerk R (2009) Can the failing Western medical paradigm be shifted using the principle of sustainabilty? Australas CollE Nutr Environ Med J 28(3):4–10

Willcox BJ, Willcox DC, Ferrucci L (2008) Secrets of healthy aging and longevity from exceptional survivors around the globe: lessons from octogenarians to supercentenarians. J Gerontol A Biol Sci Med Sci 63A(11):1181–1185

Willcox BJ, Willcox DC, Suzuki M (2013) The Okinawa way. Penguin Books, New York

World Health Organiszation (1946) Preamble to the constitution of the World Health Organization as adopted by the International Health Conference. World Health Organisation, New York